The Unpopular Truth *about Electricity and the Future of Energy*

Schernikau and Smith

The Unpopular Truth
about Electricity and the Future of Energy

An introduction to electricity in modern energy systems, including cost of electricity, energy returns (eROI), consequences of the energy transition, and recommendations for a sustainable energy policy

© Schernikau and Smith 2023

Energeia Publishing
Berlin, Germany
An der Wuhlheide 232
12459 Berlin
www.energeia-publishing.com
info@energeiapub.com

Typesetting:
Werneburg Internet Marketing und Publikations-Service
www.werneburg-im-ps.de

About the authors:

Dr. Lars Schernikau is an energy economist, entrepreneur, and commodity trader in the energy raw materials industry based in Switzerland and Singapore.

Prof. William Hayden Smith is Professor of Earth and Planetary Sciences at the McDonnell Center for the Space Sciences at Washington University, St. Louis, MO, USA.

Dedication: We dedicate this book to all mining and energy professionals around the world. These dedicated workers and their families provide us with a key building block of our modern existence: energy. Without the energy they provide, our lives would be shorter, our existence poorer and unhealthier, and we would have less time to focus on the altruistic development of humanity.

Acknowledgments: This book draws on the peer-reviewed scientific paper *Full cost of electricity 'FCOE' and energy returns 'eROI'*, authored by Dr. Lars Schernikau, Prof. William Hayden Smith, and Prof. Rosemary Falcon, published in the Journal of Management and Sustainability Vol. 12, No. 1, June 2022 issue at the Canadian Center of Science and Education in 2022 (Schernikau et al. 2022). In addition, we appreciate the influential work and input of Dr. Bruce Everett, Kip Hansen, Prof. Howard "Cork" Hayden, Capt. Todd "Ike" Kiefer, Dr. Sebastian Luening, Dr. Wallace Manheimer, Mark Mills, Prof. Vaclav Smil, Dr. Klaus Taegder, Prof. Holger Watter, Dr. Bodo Wolf, and many more.

Schernikau et al. 2022
Full cost of electricity 'FCOE' and energy returns 'eROI'
https://papers.ssrn.com/sol3/papers.cfm?abstract_id=4000800

Table of Contents

List of Figures..10
Foreword by Bruce Everett..13
Foreword by Lars Schernikau..17
Preface and Abbreviations...21
1. Electricity and investments – the current situation......................25
2. Variable "renewable" energy and storage....................................33
    2.1. Wind and solar – the disconnect between installed capacity and generated electricity..33
    2.2. Natural capacity factors..36
    2.3. Transmission, distribution, conditioning, and black start......42
    2.4. Energy storage..46
    2.5. Hydrogen and how it compares to hydrocarbons.................54
    2.6. Material input and embodied energy....................................61
    2.7. Primary energy and heat pumps...70
3. Cost of electricity and eROI...79
    3.1. Full cost of electricity – FCOE...82
    3.2. Energy return on energy invested – eROI............................90
    3.3. The 2nd Law of Thermodynamics' impact on energy systems... 96
4. The projected future of energy and "decarbonization"..............101
    4.1. Primary energy (PE) growth until 2050.............................102
    4.2. Energy shortages and their impact on prices and economic activity 105
    4.3. Decarbonization and "Net-Zero"..111
5. The realistic future of energy and sustainability........................119
Executive Summary...125
Keywords..130
References..131
Appendix 1..144
Appendix 2..146
Appendix 3..148
Appendix 4..150
Appendix 5..152

List of Figures

Figure 1: History of Earth's climate over 600 million years............................18
Figure 2: Top 10 countries – electricity generation...25
Figure 3: Overview of global primary energy and electricity..........................26
Figure 4: Investment in coal less than half of wind/solar, while coal ... .27
Figure 5: Fossil fuel capex of S&P Global 1200 energy companies ... .30
Figure 6: Global investment in energy transition...31
Figure 7: German installed power capacity, electricity production ... .34
Figure 8: Summary of shortcomings of variable "renewable" energy ... .35
Figure 9: Global wind and solar maps for Europe, Africa, and Asia..............37
Figure 10: Germany's wind generation 25 April to 10 May ... .39
Figure 11: Laws of physics limit technological improvements for wind ... .41
Figure 12: US transmission grid growth insufficient for "Net-Zero"............44
Figure 13: US reported electric disturbances by season...................................45
Figure 14: High correlation of output for hydroelectric energy ... .50
Figure 15: Illustrating energy densities – battery versus coal.........................52
Figure 16: Volumetric energy densities of hydrogen, gasoline, and ... .57
Figure 17: Arrival of the first liquid H_2 carrier in Australia in 2022...............59
Figure 18: Hydrogen electrolysis and methanization (illustration)...............61
Figure 19: Global grown and mined material extractions from 1970-2017..62
Figure 20: Embodied energy of selected construction and ... .63
Figure 21: Share of top three producing countries in production of ... .65
Figure 22: Base-material input per 1 TW generation...66
Figure 23: Comparing mineral needs for "renewable" technologies..............67
Figure 24: Mineral demand for "clean" energy technologies ... .68
Figure 25: Physics of a heat pump..75
Figure 26: Heat pump performance reduces as it gets colder.........................77
Figure 27: IEA's misleading LCOE comparison of intermittent solar ... .81
Figure 28: Full cost of electricity (FCOE) to society – a complete picture...88
Figure 29: Advanced societies require high net energy efficiencies (eROI).91
Figure 30: The concepts of eROI and material efficiency – illustrative.........93
Figure 31: Example of proposed eROI study to compare coal and solar ... 95
Figure 32: 1st and 2nd Law of Thermodynamics in closed systems..............96
Figure 33: Current and future energy security in China..................................98
Figure 34: Wind/solar capacity forecast for 2050 to be almost 4x ... .102
Figure 35: Global primary energy from 1750 to 2050......................................104

Figure 36: Urban heat island effect – NASA's "ECOSTRESS"........................113
Figure 37: Global primary energy from "carbon-free" sources 1965-2100..115
Figure 38: Household income spent on energy by total household income 116
Figure 39: Environmental impact of energy systems ... .120
Figure 40: Variable "renewable" energy does not fulfil objectives ... .121
Figure 41: Sustainable energy policy and the New Energy Revolution......123
Figure 42: The Thermodynamic System C-H-O-N..144
Figure 43: Fuels in a Thermodynamic System $C-H_2-O_2$; Conversion ... .145
Figure 44: Conversion of Energy Units...144

Foreword by Bruce Everett

For the last 50 years, the elimination of fossil fuels – oil, natural gas and coal – has been a solution in search of a problem. During the 1960s, urban air pollution was the worry. During the 1970s and 1980s, the national security threat from imported oil was the primary driver of energy policy. The depletion of a limited resource base was a constant concern, culminating in the "peak oil" movement of the 1990s and early 2000s. Today, "catastrophic" climate change has taken center stage as the main reason to seek an accelerated elimination of fossil fuels.

Despite this constant anxiety, numerous policy initiatives and the expenditure of trillions of dollars on alternatives, fossil fuels remain the dominant source of energy throughout the world, and their use continues to grow in absolute terms. To understand the reasons for this seeming contradiction, Dr. Lars Schernikau and Prof. William Hayden Smith have compiled a complete overview of how the global energy economy actually works, as opposed to the way it is presented in the popular media.

This volume is based on three principles critical to understanding energy.

The ***first principle*** is a focus on human well-being as the cornerstone of any policy analysis. Energy is the lifeblood of modern economies. One of the great historical accomplishments of the 20th and early 21st centuries has been improved living standards in industrialized countries accompanied by an extraordinary reduction in poverty in the developing world. Fossil fuels and nuclear energy have played a major role in this achievement. The continuation of global progress toward the elimination of poverty and the improvement of general living standards requires recentering the energy debate on human welfare as opposed to a myopic determination to reduce fossil fuel use.

The ***second principle*** is a recognition that energy choices are constrained by the laws of thermodynamics, chemistry, geography, meteorology, and economics. Ignoring these constraints can lead to the waste of large amounts of scarce capital, lower living standards in industrialized countries, a threat to the process of poverty reduction in the developing world, and undesired environmental effects. Understanding these real-world constraints, on the other hand, helps to explain why the global energy economy has developed as it has.

Electricity, which is a central and growing element in the world energy economy, is a particularly complex problem. Gasoline, for example, is easy to store, and ample inventories are held in refineries, terminals, filling stations, and automobile fuel tanks. Electricity, on the other hand, is difficult and expensive to store, requiring power companies to generate exactly the required amount of electricity, not only hour by hour but second by second.

The popular narrative is that the sun and wind provide unlimited free energy that requires only political will to harness. In reality, the laws of thermodynamics limit the amount of wind and solar energy that can be captured using known technologies. These technologies generate electricity only when nature provides it, not when consumers need it. Conventional power plants, on the other hand, can generate power as needed. The lack of storage capability makes it impossible to manage an electric power grid without a substantial component of fossil fuel and nuclear capacity. Moreover, the equipment required to convert the "free" wind and solar energy into usable electricity is very expensive.

The ***third principle*** is that the evaluation of energy options requires a review of the complete supply chain from beginning to end. Automobile companies, for example, are fond of saying that electric cars are "zero emission". This characterization, however, ignores the energy, labor, materials, and logistics required to generate the electricity that charges the vehicle's battery. It also ignores the energy, labor, materials, and logistics required to build, assemble, and ultimately dispose of the electric vehicle itself, with its large battery. When the full "life cycle" is considered, energy choices often look quite different.

In articulating these three principles, Lars and Bill have scrupulously avoided taking partisan positions and are offering information that will be important and useful to everyone, regardless of political outlook. As a result, they have produced an invaluable reference work that should be on the bookshelf of anyone interested in energy policy, electricity markets, and environmental protection.

Bruce McKenzie Everett, PhD

Foreword by Bruce Everett

Bruce Everett is an energy economist with nearly 50 years of experience in the international energy industry. He received a BA from Princeton University in 1969 and a PhD from The Fletcher School in 1980. Between 1974 and 1980, he served in the Federal Energy Administration and the US Department of Energy in the Office of International Affairs. He joined ExxonMobil Corporation in 1980 and held a variety of executive positions all over the world in corporate planning, oil, natural gas, coal, electric power, business development, and government relations. After retiring from ExxonMobil in 2002, he taught economics as Adjunct Professor at the Georgetown University School of Foreign Service and as Adjunct Associate Professor at The Fletcher School at Tufts University. He currently serves on the Board of Directors of the CO_2 Coalition (https://www.co2coalition.org).

Foreword by Lars Schernikau

Energy is all-encompassing, the basis of everything around us. As obvious and simple as it may be, it took me decades to truly internalize this. I grew up with energy and raw materials, from the day I opened my eyes, but started working with raw materials and energy only 20 years ago when I became a commodity trader. During that time, I spent a large portion of my professional career in the global coal markets. However, professionally I also started to deal with ore products such as iron ore, lithium ore, copper ore, chrome ore, and much more.

This clearly makes me biased in writing about electricity and writing critically about variable renewable energy. However, I would like you to consider that the fossil fuel business will thrive when energy shortages prevail. Because energy-starved times will always be accompanied by high raw material and power prices, leading to extra profits for anyone producing or even trading energy raw materials and generating electricity. This can be seen in the record profits earned by large oil, gas, and coal companies during 2021 and 2022. Thus, if anything, I am negatively incentivized to write about how the world can avoid energy shortages. I should, from a private economic point of view, keep quiet. However, this book is about what is right for the world and what we can do to optimize our energy production.

Reliable and affordable access to energy should never be political. Unfortunately, energy has been misused by both sides of the political spectrum for exactly that, political agendas. It should be any government's interest to have a good energy mix, reduce dependencies, ensure affordability, reliability, and of course limit the environmental footprint. Unfortunately, history is full of examples of exactly the opposite. The 2022 Ukraine conflict yet again shows how intertwined energy and politics are. Remember, however, that energy shortages started in 2021, so Putin was not the cause, but made it worse and accelerated the process towards global energy starvation.

Over the long term, we need to find a solution for our energy problem. I have learned that we have hundreds of years of fossil fuels left in the ground. However, and quite obviously, we cannot dig fossil fuels up forever. Not only because there are not enough of them, but really because we will need so much more energy in the future. Oil, coal, and gas will be neither sufficient nor efficient enough to sustain our substantial thirst for the amount of energy we will need for our natural, human, and scientific evolution and development in the centuries to come. That is what sparked my interest in learning more about electricity and energy beyond coal, oil, and gas. I strive to understand what the future of energy can be and what it cannot be. I continue learning every day and ask you for forgiveness already now for any inaccuracies in language or content you may find.

Humanity has amassed more scientific knowledge since World War II than over the previous one million years of human development. Following the agricultural revolution, made possible by a drastic temperature increase during the early Holocene (Figure 1), it took 10.000 years to create civilization in Europe. It only took a century for the steam engine to facilitate the development of modern industry. The nuclear force discovered in the mid-twentieth century increased the power available to a single human by a factor of one million.

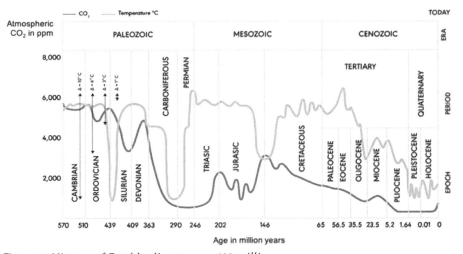

Figure 1: History of Earth's climate over 600 million years

Note: Graph of global temperature and atmospheric CO_2 concentration over the past 600 million years detailing Earth's recent eras, periods, and epochs.

Note: The Holocene started around 13.000 years ago and is what we call the current interglacial warm period. It forms part of the late Pleistocene Ice Age, which started less than two million years ago.

Source: Moore 2017 based on Nahle 2009

Scientific progress will lead us to what is often referred to as Kardashev's Type 1 civilization in which humans are able to harness all the energy available to our home planet and store it for consumption (McFadden 2019). Michio Kaku believes this may only be 100-200 years away, which seems possible. This contrasts with a Type 0 civilization – today's world, a sub-global civilization that harnesses power primarily from raw materials.

Humanity's development, however, will not be limited to scientific advances but will also include spiritual advances that will allow us to better understand the energetic connection between matter and mind. The argument is, therefore, that our "energy problem" will be solved within a century or two through the ***New Energy Revolution***, as discussed in the last chapter of this book.

Nature has evolved in ways we often forget. Dinosaurs went extinct just 65 million years ago (not even 20 minutes ago in Earth's 24-hour history). Imagine a world without flowers; is that possible? Yes, it was possible just 120 million years ago; that is approximately when the first flowers appeared. Life did not require flowers for the previous hundreds of million years. It seems that flowers have no purpose other than to provide beauty. Obviously, nature has started to take advantage of flowers by fueling biomass reproduction, but really, it was not necessary.

Flowers, along with crystals, precious stones, and birds, have held special significance for the human spirit. What else will nature provide us with? Future human development will surely include a better understanding of how the world is connected. We will learn to better understand our mind and to use this newly found power to heal and to experience unimaginable happiness. Some of the world's most famous neuroscientists encourage us to embrace meditation and spirituality. Meditation provenly activates parts of our brain that help us heal and access knowledge and connections we have not yet dreamed of.

Why do I mention this? To make you aware of how little we know about the future, other than that evolution and development are vast, fast, and surprising. **The Future of Energy** will be developed by our human mind accessing all the knowledge that we don't yet even know that we don't know it. This is where my interest lies: in starting to learn more about the things I am not yet aware of and starting to see how we can truly make a positive difference to our world by being all-encompassing rather than limiting.

I strive to see the future differently and positively by inclusion, not by exclusion, by realizing physical limits but being open to what may be. I truly hope that you will find this book valuable in better understanding how energy works, and how it does not work.

I am looking forward to your feedback about this book.

Lars Schernikau

Dr. Lars Schernikau is energy economist, entrepreneur, commodity trader, and author. Educated at NYU in the US, INSEAD in France, and TU Berlin in Germany, he has worked with commodities for two decades in Asia, Europe, Africa, and North America with focus on coal and ore products. Previously, he worked for the Boston Consulting Group in the US and Germany.

Preface and Abbreviations

Energy (in Watt-hour or Wh, in German "Arbeit oder Energie") versus ***Power*** (in Watt or W, in German "Leistung").
- Energy is the capacity to do work. Power is energy extended per unit of time. Energy can be transformed from one form to another.
- Once you know both the energy storage capacity (i.e., MWh) of a battery and its output power (i.e., MW), you can divide these numbers to find how long the battery will last.
- Energy is stored in a Tesla battery (i.e., 100 kWh). To move the car, power (kW), or expenditure of energy, is required. The rate at which energy is expended tells how fast the car can accelerate. The battery, "filled" with energy (kWh), is drained over time depending on how much power (kW, also "horsepower") is required for moving the car, which – in turn – depends on how you drive and the surrounding conditions.

Capacity Factor "CF" (in German "Nutzungsgrad") is the nature derived percentage of power output achieved from the installed capacity for a given site, usually stated on an annual basis. The natural capacity factor does not depend on technology or utilization, but is only driven by natural conditions, such as sunlight hours, wind-hours, or availability of water in a river.
- The natural capacity factor is site-specific and cannot be changed with technology. Thus, capacity factor here is only driven by natural conditions, not by technology or operations (or "utilization"). In other words, when technology fails, or a power plant is turned off on purpose, this will reduce the utilization, but not the natural driven capacity factor.
- Capacity factor is different from the common conversion efficiency factor. For comparison, conversion efficiency measures the percentage of input energy transformed to usable or output energy.
- In Germany, photovoltaics ("PV") achieve an average annual natural capacity factor of ~10-11%, while California reaches an annual average CF of ~25% (Schernikau and Smith 2021). Thus, California yields almost 2,5x the output of an identical PV plant in Germany.

- It is important to distinguish between the average annual capacity factor and the monthly, or better weekly, daily, or even instantaneous natural capacity factor, which is relevant when keeping an electricity system stable which requires demand to always equal supply, every second.

Electricity versus other uses of energy. Primary energy is used to generate electricity (often referred to as power), for transportation needs, heating requirements, and diverse industrial purposes. This book focuses on electricity or "power" markets. To avoid confusion, we shall use "electricity" rather than "power". Wherever power is used, it shall equate electricity.

Conservation of energy – the 1st Law of Thermodynamics essentially states that energy can neither be created from nothing nor lost into nothing; it can only be converted from one form to another. Different forms of energy include thermal, mechanical, electrical, chemical, nuclear, and radiant energy.

Entropy of Energy – the 2nd Law of Thermodynamics distinguishes between useful energy (low entropy), which can perform work, and less useful energy (high entropy), which cannot easily perform work. The use of energy to do work degrades the energy quality. For example, passing electricity through a resistance heater "degrades" the electricity to radiant heat or, in other words, warm air.

- Entropy is a measure of randomness or disorder within an energy system, where greater disorder equals greater entropy.
- Whenever energy is converted from one form to another, there is always a fraction of energy that becomes useless or waste energy (entropy/disorder increases).
- For example, sunlight converted for transmission to the consumer and finally converted into electrochemical energy stored in a battery suffers losses at every step, meaning that much less than 1% of the original solar energy ends up stored in the battery. The stored electrochemical energy is then converted to electricity that turns an electric motor; this turns the wheels of a car to move it along a road. The reported EV miles per gallon equivalent ignores all steps but the last two because sunlight is "free". The collection, transmission, and conversion systems are, however, not free and

obey the 2nd Law of Thermodynamics. Thus, the more complex energy processes are, the more useful energy is lost.
- As Planck put it: "Every process occurring in nature always increases the sum of the entropies of all bodies taking part in the process, at the limit – for reversible processes – the sum remains unchanged". The 2nd Law of Thermodynamics thus explains why perpetual motion machines are not possible.

AC – Alternating current
CCUS – Carbon capture utilization and storage
CSP – Concentrated solar power
DC – Direct current
eROI – Energy return on energy invested, or energy returns
EV – Electric vehicle
HELE – High-efficiency, low-emission
IEA – International Energy Agency in Paris
FCOE – Full cost of electricity
LCOE – Levelized cost of electricity
LDES – Long-duration energy storage
MIPS – Material input per unit of service
PE – Primary energy (PES = primary energy supply)
PV – Photovoltaic
PtHtP – Power-to-Heat-to-Power storage technologies, such as molten salt
USC – Ultra-supercritical
VRE – Variable renewable energy, such as wind and solar
WT – Wind turbine
~ – Approximately

Embodied Energy – the accumulated energy required to produce for example steel or aluminum. Embodied energy or embedded energy is therefore a component of any material or product.

1. Electricity and investments – the current situation

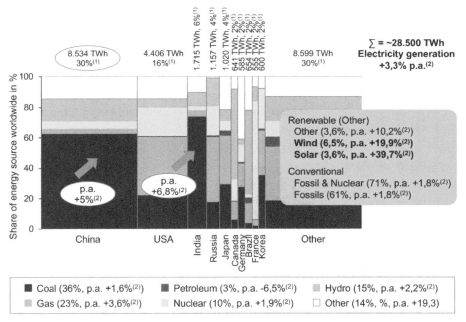

Figure 2: Top 10 countries – electricity generation
(1) Electricity production share of respective country; (2) CAGR – compound annual growth rate 2012-2021 in %.

Source: Schernikau Research and Analysis based on BP Statistical Review of World Energy 2022 (BP 2022)

Fossil fuels – in order of importance *oil, coal, and gas make up ~80% of global primary energy ("PE") production*, totaling ~170.000 TWh or ~600 EJ. Despite Covid-19, geopolitical turmoil, and significant wind and solar capacity additions, it is estimated that the percentage will not change in 2022; quite to the contrary, coal is making a comeback (IEA 2022). Coal and gas made up ~60% of global gross electricity production, totaling ~28.400 TWh in 2021. Three countries alone, China, the USA, and India, make up ~50% of global electricity consumption (Figure 2). It is important

25

to note that global electricity production consumes up to ~40% of primary energy, with transportation, heating, and industry accounting for the remaining ~60% (Figure 3 and Figure 35, p 102).

Realistic primary energy growth of ~50% until 2050 is driven by ~25% population growth and ~20% average per capita energy demand growth. This contrasts with IRENA's, McKinsey's, and the IEA's "Net-Zero" pathways, which often assume much less growth and either flat or even a ~10% drop in primary energy. McKinsey 2022b assumes a 14% rise until 2035 and thereafter flat primary energy consumption until 2050. IEA Net-Zero 2021 assumes a 10% drop in primary energy consumption demand as early as 2030. The "primary energy" measure and the misconception of falling primary energy with solar and wind's increased penetration is discussed in Chapter 2.7.

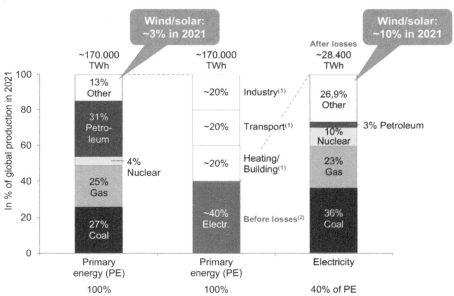

Figure 3: Overview of global primary energy and electricity

(1) Only the portion of Industry/Transport/Building that is not included under electricity; (2) assumed worldwide net efficiency of ~33% for nuclear, ~37% for coal, ~42% for gas. With an assumed avg. ~40% efficiency => 28.400 TWh becomes 71.000 TWh or roughly 40% of ~170.000 TWh.

Source: Schernikau Research and Analysis based on BP 2022, IEA Statistics 2021

1. ELECTRICITY AND INVESTMENTS – THE CURRENT SITUATION

Current energy policy focuses on the electrification of energy, thus significantly increasing electricity's share of primary energy by increasing the use of electricity for transportation (see EVs), heating (see heat pumps), and industry (see DRI, producing steel using hydrogen). Therefore, this book focuses on electricity. For a more comprehensive discussion on transportation, we recommend Kiefer's 2013 *Twenty-First Century Snake Oil*, which includes details on hydrocarbons and biofuels for transportation that are not covered herein in detail.

Despite the trillions of US dollars spent globally on the "energy transition" (Figure 4), the proportion of fossil fuels as part of total energy supply has been essentially constant at around 80% since the 1970s, when gross energy consumption was less than half as high (WEF 2020). Also in Europe, fossil fuels' share is still above 70%. Kober et al. 2020 among others, confirm that total primary energy consumption more than doubled in the 40 years between 1978 to 2018. At the same time, the energy intensity of GDP improved by a little less than 1%, confirming the Jevon's Paradox that ***energy efficiency improvements are in principle offset by higher energy demand*** (Polimeni et al. 2015).

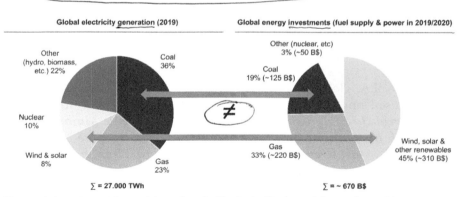

Figure 4: Investment in coal less than half of wind/solar, while coal provides ~4x more power

Note: Chart on the right includes investment in fuel supply and power; for gas, it is

assumed that 40% of total "oil & gas" fuel supply investment went into gas (511 B$ x ~45% = 220 B$).
Source: Schernikau Research and Analysis based on IEA and BNEF Data; IEA World Energy Investment 2020

Variable "renewables"[1] in the form of wind and solar accounted for ~3% of global primary energy and ~10% of global gross electricity production in 2021 (refer to Schernikau and Smith 2021 for more details on solar PV and Smith and Schernikau 2022 on wind). Thus, fossil fuels still exceeded wind and solar by a "Fossil to Wind-Solar Factor" of 27x for primary energy and 8x for electrical power production (IEA Statistics 2021). It must be noted that thermal electricity generation also produces much-needed heat as a byproduct – through co-generation – that is used for industrial and household purposes. This heat generated would have to be replaced, and would further increase electricity demand, if the world were to reduce thermal power generation.

Note: Other forms of energy supply that are categorized as "renewables" or "low carbon"– such as nuclear, biomass, hydro, geothermal, or tidal power – are not discussed further in this book as they are not considered erratic or variable and can, in principle but with some exceptions, be included in dispatchable energy resources. Biomass has a very low net energy efficiency, is limited by suitable cropland, but consistently provides around 7% of global primary energy. Hydro energy is very energy efficient but is limited by suitable natural river flows, and it consistently provides around 3% of global primary energy and 7% of electricity. Nuclear is the most energy-efficient way of generating electricity and contributes around 3% of global primary energy and 10% of electricity. For more details on biofuels, please refer to the detailed plain-language summary provided by Kiefer 2013 as well as EPA 2022.

As per the IEA, global energy and fuel supply investments are increasingly focused on "low carbon" energy sources. **Wind and solar – for their currently insignificant energy share – receive almost half of the total investment that oil, coal, and gas receive combined (Figure 4). Moreover, this ratio continues to change quickly in favor of** wind and solar as investment in fossil fuels continues to fall while investment in wind and solar soars. Subsidies for both "renewables"

---

[1] We write "green" and "renewables" in quotation marks because wind and solar are neither really green nor really renewable. The fuel they use, i.e., wind and solar, is green or renewable, but wind and solar systems in their entirety and their impact on the environment are neither green nor renewable, as detailed in this book. When we refer to "renewables" without specifying further, we mean what are commonly considered to be "renewables", such as wind, solar, biomass, geothermal, and hydro power.

and conventional energy are not considered here[2], although the EU spends more subsidies on "renewables" than on fossil fuels, even in absolute terms (EC 2022, p30).

JP Morgan 2022 summarized the falling capex of S&Ps global 1.200 energy firms while absolute fossil fuel energy demand continues to increase (Figure 5). Dr. Fatih Birol, IEA's executive director and one of the world's foremost energy economists, told the Guardian in May 2021 (Harvey 2021): *"If governments are serious about the climate crisis, there can be no new investments in oil, gas and coal, from now – from this year".*

Additional important global energy statistics for the pre-Covid year 2019 include the following (BNEF 2021, IEA Investments 2020, IEA WEO 2020, IEA Investments 2021):

- Global installed electrical power generation capacity from coal, gas, and oil equaled 57% of all installed capacity in 2019. This capacity generated ~62% of global power.
- Global installed wind and solar power generation capacity equaled 18% of all installed capacity. This wind and solar capacity generated ~8% of global power (2021 it was ~10%).
- Global investment in oil, coal, and gas supply and power generation equaled US$ ~700 billion in 2020, of which US$ ~100 billion was invested in fossil fuel power generation (down from US$ ~180

---

[2] The IMF reported around US$ 450 billion of global "explicit" fossil fuel subsidies in 2020 and around US$ 5,5 trillion in so-called "implicit" subsidies for fossil fuels (IMF 2021). IRENA estimates that "renewables" received around US$ 130 billion of subsidies in 2017 (IRENA 2020), thus per MWh significantly more than fossil fuels. The EU already spends more subsidies on "renewables" than on fossil fuels in absolute terms (EC 2022, p30). Everett 2021 discusses the Social Cost of Carbon in detail and concludes: *"Leaving aside its scientific and economic uncertainties, the government's Social Cost of Carbon is so sensitive to input assumptions that small, quite reasonable variations can produce almost any price you wish. As a result, it is not a suitable tool for guiding public policy, including taxes on energy."* Therefore, we dismiss the concept of "implicit subsidies" as virtually any number can be calculated depending on the assumptions made, and all forms of energy receive "implicit subsidies", whether it be solar, wind, biomass, hydro, gas, coal, or nuclear. For example, wind and solar are not CO_2-taxed even though their production and recycling cause significant GHG emissions. For projected costs of global warming, please refer to Nordhaus 2018, Lomborg 2020, and Kahn 2021. To compare subsidies correctly, they will always have to be baselined on a per unit of output energy basis, which is rarely done.

billion ten years ago); the remainder was used for securing fossil fuel supply.
- The IEA's Sustainable Development Scenario (SDS) requires this investment to increase annually to US$ 750-800 billion in the period 2025 to 2030 to keep the energy system supplied and stable.
- Global investment in wind and solar alone reached US$ ~370 billion in 2021. Total investment in the "energy transition", which also includes investment in electricity consumption such as electric vehicles (EVs), biofuels, storage, CCS, and hydrogen, was US$ ~700 billion. Investment in VRE exceeded US$ 4 trillion in the past 20 years (see Figure 6).
- The red line in Figure 6 illustrates that only a very small expenditure increase in "renewable" energy production occurred after 2011. Almost all increased expenditure was for electricity consumption, rather than production, which is often referred to as "investment in the energy transition".

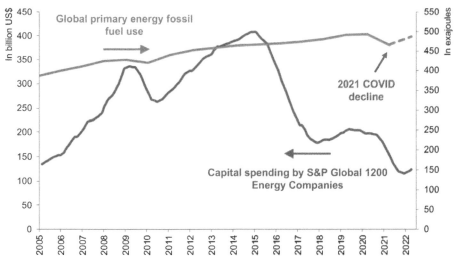

Figure 5: Fossil fuel capex of S&P Global 1200 energy companies slumps despite growing demand

Note: Dotted line indicates estimates.

Source: JP Morgan 2022, p6, based on BP, Bloomberg, IEA, and JPM analysis

1. ELECTRICITY AND INVESTMENTS – THE CURRENT SITUATION

It should be noted that despite this *investment disparity – US$ ~370 billion for ~3% of primary energy versus US$ ~700 billion for ~80% of primary energy*, a relative investment factor per unit of energy supply of ~10:1 to ~15:1 (Figure 4) – the IEA confirmed in July 2021 that *"...[renewables] are expected to be able to serve only around half of the projected growth in global [electricity] demand in 2021 and 2022"* (IEA Electricity 2021, p3). In fact, the IEA confirmed in January 2022 that over 2/3 of electricity growth in 2021 came from conventional fuels, including over half from coal alone (IEA Electricity 2022). For the foreseeable future, increases in "renewable" generation will only make up a fraction of primary energy. This is driven by the fact that currently only around (40%) of primary energy is used for electricity generation, with the remainder being used for industry, transportation, and heating (Figure 3, and Chapter 2.7 on primary energy and Footnote 4).

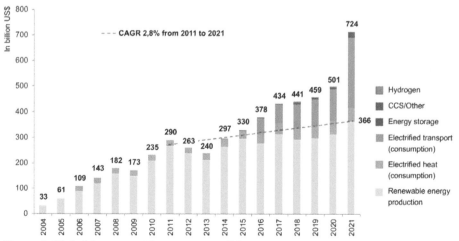

Figure 6: Global investment in energy transition

Note: The red line illustrates the relatively small investment increase in "renewable" electricity production. Investment in "renewable" electricity consumption now equals investment in production.

Source: BNEF 2022: Energy Transition Investment Trends (https://bit.ly/3e4uYXj)

2. Variable "renewable" energy and storage

2.1. Wind and solar – the disconnect between installed capacity and generated electricity

Germany is the leading industrialized nation in the move toward decarbonization and has officially invested EUR ~400 billion in the "energy transition" since 2000[3], reducing the share of nuclear and fossil fuels (BfWE 2020). The actual number is larger, driven by the cost of idle thermal power plants, the cost of grid prioritization, and other factors. It should be noted that nuclear is the most energy efficient (see Chapter 3 on eROI) and least polluting way of generating electricity, but it faces other challenges that are not further detailed herein. However, as Europe has been reducing its production of fossil fuels, the continent's dependence on energy raw material imports, mostly from Russia, has increased significantly over the past two decades. This policy appeared to support fossil fuel production, as long as it was not located in Western Europe.

With the money invested in the "energy transition" – up to 2021 – Germany had reached a wind/solar share for gross electricity production of ~28%. The primary energy share of wind and solar[4], however, was still only 5% (please refer to Chapter 2.7 for a discussion on primary energy and "renewables"). To achieve this "transition", Germany's installed power capacity had to double (Figure 7). Consequently, the "renewable" energy sector has grossly underperformed compared to the investment in real energy terms, and Germany's electricity prices have become the highest in the G20. This underperformance, however, is due to the low natural capacity factor, low energy

---

3 Note: The numbers include only "EEG-Gesamtvergütung" (EEG compensation package), but no other investments, research, subsidies, etc.
4 The fall in primary energy is due to, among other factors, the assumed 100% efficiency of wind and solar electricity when calculating their share in PE. In other words, it is mistakenly assumed that wind and solar electricity generation was converted at 100% efficiency without any losses or energy costs (see Chapter 2.7 for more details). If one were to assume a more realistic lower net efficiency, the primary energy share of wind and solar would increase and total primary energy would be higher.

efficiency, and the other inherent shortcomings of variable "renewable" energy discussed herein (Figure 8), rather than bad implementation or intentions.

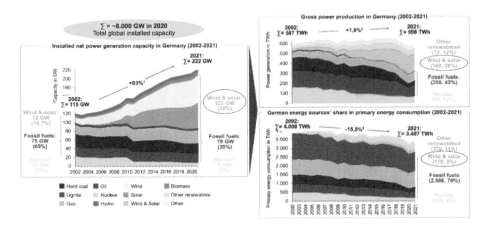

Figure 7: German installed power capacity, electricity production, and primary energy
Notes: (1) CAGR: +3,5%; (2) CAGR: +0,1%; (3) CAGR: -0,9%; (4) Including hydro and biomass.
Source: Schernikau Research and Analysis based on Fraunhofer 2022, AGE 2021, Agora 2022, see Footnote 4

During the 20 years from 2002 to 2021, Germany's installed power capacity almost doubled from 115 GW to 222 GW, while total electricity consumption was essentially flat and primary energy[4] fell over 15% (Figure 7). Over the decades to come, Germany expects a significant increase in electricity consumption due to hydrogen demand and the electrification of transportation, heating, and industrial processes to satisfy increased demand from consumers and industry, as required by the German "Energiewende".

The global average looks slightly better. Of the total 2020 global installed energy capacity of ~8.000 GW or 8 TW (Figure 10), around 18% or ~1.400 GW was wind and solar, which contributed ~8% to global electricity and ~3% to primary energy (BP 2021, IEA Power 2019, IEA Statistics 2021). After the installation of almost 200 GW of solar PV in 2021, the world celebrated the first 1 TW of installed solar capacity in March 2022 (PV-Mag 2022).

Figure 7 illustrates the substantial disconnect between installed capacity and generated electricity. It appears that in countries such as Germany, given the average natural capacity factors for wind and solar, a doubling in installed capacity will lead to less than 1/3 of electricity supply and a contribution of less than 10% to primary energy. The reasons for this disconnect are manifold and impact the world of electricity in many ways. Figure 8 lists the "shortcomings" of variable "renewable" energy (VRE) in the form of wind and solar for electricity generation that explain the apparent disconnect. These VRE deficiencies can only be partially reduced through technological improvements.

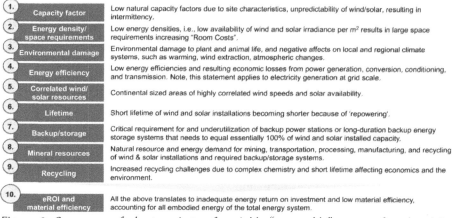

Figure 8: Summary of shortcomings of variable "renewable" energy for electricity generation
Source: Schernikau Research and Analysis

Despite the sun's immense power, the energy available per m² from natural wind and solar resources is limited, intermittent, and too small to allow efficient electricity generation at grid scale (low energy density). The additional negative effects of wind and solar on vegetation, local and regional climate, animal life, seaways, bird flyways, and bird, bat, and even insect populations also must be considered. These effects originate primarily from the required large land area (Schernikau and Smith 2021, Smith and Schernikau 2022).

Technological advances will further increase the net efficiencies of wind and solar installations. However, physical boundaries, as described by the Betz Law and Schockler-Queisser Limit (see Figure 11), dismiss the possibility of ten-fold improvements. There is no prospect of a paradigm shift in energy from PV or wind as, for example, is promised for quantum computing. One cannot compare technological advances in energy with computing as they follow different laws (Figure 32, p94).

The 33% quantum efficiency Schockler-Queisser limit for silicon can be exceeded with multilayer PVs, which so far are unstable and less durable than silicon PV panels. Today, they already surpass mono-crystalline silicon's quantum efficiency by around 50%, but 20-year operational lifespans for multilayer PVs are not within reach. Technological improvements and new materials, such as perovskites and quantum dots, may overcome the stability and durability problems in time, but 100% quantum efficiency is the absolute physical maximum and will never be reached.

Thus, *technological improvements may increase PV's quantum efficiency by a factor of two, but not by the multiple required to compete with conventional energy generation and surpass the required eROI hurdle* at grid scale. Let us be clear that conventional energy generation also improves its efficiency over time. For example, the latest USC HELE coal-fired power plants in Japan and Germany attain over 60% efficiency and achieve the required reliable and dispatchable energy output while also cutting their fossil fuel needs and GHG emissions, as well as practically eliminating pollutant emissions. This appears to be a more effective means to control emissions than has been achieved so far with "renewable energy" resources (see below).

2.2. Natural capacity factors

Capacity factors (see preface) are the ratio of maximum possible output of the energy converter, including all its physical limitations, to the output actually achieved under the conditions of the site, assuming no operational or technological failures or outages.

- For a photovoltaic (PV) park, the natural capacity factor depends entirely on the intensity and duration of the sunlight, which is affected by seasonality and cloudiness, day and night, and the ability to maintain the PV panel surface's transparency under local ambient conditions, e.g., dust in the Sahara.
- Wind turbine farm natural capacity factors depend on the site's wind speed distribution and the saturation speed of the wind turbine. The capacity factor of a wind turbine is determined by the number of hours per year in which the wind farm operates at or above the saturation wind speed (Smith and Schernikau 2022). If

the design wind saturation speed is set low, e.g., 4-5 m/s, the wind farm produces little energy, even for a high capacity factor. Typically, wind saturation speeds are 12-15 m/s.

When we speak of the natural capacity factor here, we are only referring to the **nature-derived** capacity factor, not the technological or **operationally driven** capacity factor (let us call this "**utilization**", it is also often referred to as plant load factor, PLF). In other words, when technology fails, or a power plant is turned off on purpose, this will reduce the utilization but not the natural capacity factor. Thus, one would have to multiply the nature-derived capacity factor by the technological or operational utilization to obtain the "**net load factor**". The energy press has recently pointed out that coal or gas have capacity factors of 60% or less on average. However, such a number is not the nature-derived capacity factor; it is the net load factor and as such declines with higher penetration of wind and solar and contributes to power system cost increases. Conventional power plants have near 100% natural capacity factors, but their operational and technological utilization often falls significantly below 90%.

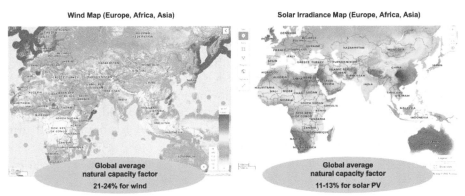

Figure 9: Global wind and solar maps for Europe, Africa, and Asia

Source: Global Wind Atlas 2022 (setting Mean Power Density – for 10% windiest selected regions at 100m height), Global Solar Atlas 2022 (setting Direct Normal Irradiance, DNI)

Needless to say, the natural capacity factor of wind and solar cannot be predicted or guaranteed for any given time frame. The natural capacity factor can be estimated on an annual basis but still varies widely year by year (see Europe 2021). ***Thus, natural capacity factors worldwide are a direct result of the location of the wind or solar installation; they do not in any way depend on and cannot be influenced by the technology employed.*** Figure 9 illustrates how unevenly wind and solar resources are distributed across the planet. The high population of the Asian and equatorial regions has few usable wind resources. Solar resources are limited by day and night, sunshine hours in the north and south of the planet, and rainy seasons/monsoons in the equatorial regions. Australia, South Africa, the Sahara, and California are the regions with the highest solar irradiance.

Capacity factors in Europe tend to be higher for wind than for solar. Wind installations in Northern Europe may reach an average of over 30%, but less than 15% in India and less than 8% in Indonesia. Average, and the emphasis is on average, annual solar PV capacity factors reach around ~10-11% in Germany, ~17% in Spain, ~25% in California, and may reach 14-19% in India, but they reach less than 15% in Indonesia's populated areas (Chen et al. 2019, Fig. 2). Carbajales-Dale et al. 2014 confirm higher capacity factors for wind than for solar; they estimate global average wind capacity factors to be around 21-24% and solar around 11-13% (Figure 9).

As mentioned, the natural capacity factor is due to the site, not the PV. Thus, even a perfect PV material still needs to deal with natural capacity factors with an annual average of 10-25%, not counting for other losses mentioned above (Schernikau and Smith 2021). Storage in the form of hydrogen and transportation of energy derived from H_2 is forecast to overcome the issue of intermittency by moving excess wind and solar energy stored in the form of "green" hydrogen from "sun and wind rich regions" in the world to other regions in need. In our view, this is unrealistic at grid scale, as explained in Chapters 2.4 and 2.5.

To provide a further example regarding capacity factor: A 25% natural capacity factor means that a wind park with an installed capacity of 1.000 MW (or 1 GW) would return 250 MW of power on average over one year, assuming no technological or operationally driven outages, that is, assuming 100% utilization. It cannot be determined exactly when this electricity would be generated, and there would be hours and days, sometimes weeks, with virtually zero generation from such a wind park. Figure 10 illustrates a two-week wind lull period in Germany during April and May 2022, when this chapter was written, with less than 5% wind capacity factor on average. Conversely, persistently high winds create a large amount of excess power which cannot be tolerated on a grid system and may require the use of wind turbines to be curtailed, meaning they would produce no energy at all. There is a high correlation of wind speeds across continental and even hemispherical regions, meaning that high winds would create excess power across the entire grid, which cannot be tolerated easily either (Smith and Schernikau 2022).

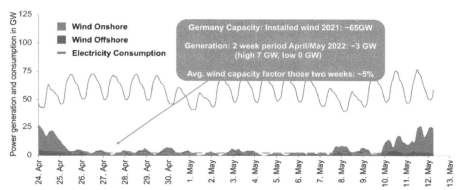

Figure 10: Germany's wind generation 25 April to 10 May during 2-week wind lull
Source: Agora 2022

The average 250 MW power generation from our theoretical wind park multiplied by the annual total of ~8.760 hours translates to 2,2 GWh of produced usable electricity for the year if we make the unrealistic assumption that this electricity can be used without other conversion, conditioning, or transmission losses (Chapter 2.7). Conversion, conditioning, and transmission are less energy efficient for wind and solar because of the spatial mismatch of demand with generation, their large variability, and the AC/DC conversion. For comparison, a 1.000 MW gas-fired power station running at an average 90% utilization or plant load factor would generate quadruple the electricity: around 7,8 GWh during that same year assuming that it is not turned off. As long as gas or coal are available and no technical breakdown occurs, the gas- or coal-fired generation can be exactly pre-determined on an hourly or even minute-by-minute basis. The IEA refers to this as the "flexibility value" in its VALCOE cost measure (see Chapter 3.1).

Given near 100% nature-derived capacity factors, conventional thermal power plants have 90+% net load factors, as do nuclear power plants when used correctly, i.e., as base power. The actual achieved load factor can be, and often is, less because of curtailments or lower utilization to match power demand and to allow the use of intermittent "renewables" (e.g., wind and solar are usually given grid priority in electricity systems, which then shuts down thermal generation). Therefore, coal, gas, and nuclear plants have capacity factors inherently 3-10x larger than variable "renewable" energy such as wind and solar. Their utilization, however, continues to decline up to a point as more wind and solar capacity enters the system. Thermal power plants with ultra-supercritical (USC) designs achieve over 60% efficiency, better even than fuel cells or any of today's wind or solar installations. Fossil fuels are thus more efficient when used appropriately and can be used all the time. Their energy values may be increased even further when modern clean combustion technologies are incorporated into high-efficiency, low-emission (HELE) power plants, thereby further increasing their environmental acceptability.

It now becomes obvious why the installed capacity needs to be much larger for wind and solar than for dispatchable power such as nuclear, wind, gas, or hydro[5]. This significant relative increase in energy generation capacity to produce the same available, but unpredictable, energy output is coupled with a significantly higher material input and energy input factor for VRE which must be offset from any fuel savings (refer to MIPS and eROI in Chapter 3).

Conversely, overbuilt VRE systems, when all weather conditions are perfect, produce far too much power, resulting in wasted energy, even when it can be stored (because storage costs energy). This overcapacity can – in principle – be partially stored, for example with the production of "green" hydrogen. However, so far even Germany does not yet produce enough excess wind and solar electricity for any as-yet non-existent long-duration energy storage systems to be

---

5   Hydro is dispatchable only as long as its impoundment is adequately filled. Norway, often considered as the EU's hydro backup, has hydro energy storage that varies by over 50% from year to year (NVE Report 2021). Below a specific impoundment level reached during drought conditions, no power can be produced. The opposite condition has been faced by the Three Gorges Yangtze River Dam in China each summer over the past two years. The impoundment has overfilled during above-average monsoons, requiring the release of water as fast as possible to avoid a collapse of the dam and a consequent flood of disastrous proportions. Thus, the capacity factor for hydro installations cannot be controlled through technology either and is weather-dependent (see Chapter 2.4), but varies less than wind and solar availability.

"charged". Overcapacity is available for only a fraction of the time during a year; it is thus an inefficient, erratic energy source to use to build up storage, and this results in low asset utilization. Presently, the small but increasingly available overcapacity is mostly unutilized, meaning that Germany has found it necessary to selectively pay neighboring countries such as Switzerland to off-take electricity in order to maintain a balanced grid and frequency in Germany. Switzerland uses the gifted electricity to "charge" its pumped hydro or for meeting its own power demand. Such payments to absorb excess electricity from Germany can have a cascading effect on the paid country (here Switzerland) in that its electricity facilities are then underutilized and idle, and therefore wasted.

- Wind → Betz Limit
  - Max 60% of kinetic energy in air that a blade can capture for conversion to electricity
  - Modern turbines do not exceed 45% conversion to electricity
  - Turbines only reach 30-45% when brand new, due to continuous degradation each year

- Photovoltaic (PV) → Schockley-Queisser Limit for monocrystalline silicon
  - Max 33% of incoming photons can be converted into electrons in silicon photovoltaic
  - State-of-the-art single-layer PV achieve over 26% conversion to electricity
  - New non-silicon options will have similar boundaries and are far from economical
  - Multi-layer PVs have reached about 45% conversion, none are as durable as silicon

- Space requirements, material input, and recycling needs of wind and solar are underestimated
  - Wind: cement, steel, balsa wood, fiber glass (for airfoils), energy and material input
  - Solar: Silicon derived from silicon oxides/quartz, silver, steel, aluminum, energy input

While we have seen large improvements in efficiencies and costs in past decade, further 10-fold improvements are impossible as we reach physical limits

Figure 11: Laws of physics limit technological improvements for wind and solar
Source: Schernikau and Smith Research and Analysis.

The nature-derived capacity factor of "renewables", which is limited by the natural availability of wind and solar, cannot be compared to utilization (also referred to as plant load factor, PLF) of a conventional power plant, which is limited by technological outages or reduced consumption requirements. It is economically incorrect to compare the net load factor or utilization of conventional power plants with nature-derived capacity factors of variable renewables.

$$\text{Nature-derived capacity factor} \times \text{utilization} = \text{net load factor} \qquad (1)$$

2.3. Transmission, distribution, conditioning, and black start

Transmission: Transmission losses are a well-known feature of modern energy systems. In conventional electricity systems, efficient larger power plants are built near demand centers, such as cities or industrial conglomerations. In "renewable" electricity systems, such optimization is not possible as wind and solar "power plants" need to be built where natural conditions are optimal. That is why transmission becomes a much larger issue for VRE. The efficient, long-distance transmission of electricity could optimize the transportation of energy from distant wind or solar farms to demand centers, as would be required by "Net-Zero" pathways, but this would not overcome the other inherent shortcomings of VRE summarized in Figure 8.

One example of a long-distance transmission solution is the proposed Euro-Asia Interconnector Project 18, which would deliver 2 TW of power from the Northern Territory of Australia to Singapore over a sea distance of 4.500 km (EuroAsia Interconnector 2017). If ever built, the project would demonstrate very long-distance submarine power transmission. **Less than 60% of the power fed into the system in Australia would be expected to reach the consumer in Singapore**, according to the design specification. That is, 2 TW would be delivered compared with a 3,3 TW source capacity. Such losses are unavoidable when "producing" a reliable, sufficient, and stable supply of electrical power between on- and offshore wind farms and consumers (Smith and Schernikau 2022).

When undersea cables fail, the EEA 2009 notes, the lines are typically out of service for months. A deep-sea cable such as the Euro-Asia Project would also be subject to turbidity currents, which have repeatedly broken communication cables between Europe and North America (ScienceDirect 2011).

Another important aspect of the inefficiency of transmitting "renewable" electricity to consumers has been highlighted by DeSantis et al. 2021. The paper compares the cost of transmitting energy using transmission lines vs. transmitting energy through pipelines in the form of oil products or gas. The cost difference is substantial, a liquid fuel pipeline may transmit energy at 0,8 to 2,2 US$/MWh/1.000 miles. While DC lines cost over 40 US$/MWh/1.000 miles, a multiple of 20 to 50x. Transporting conventional fuels using conventional methods, such as pipelines, trains, or vessels is far more efficient than transmitting energy using electric transmission lines. "*The higher cost of electrical transmission is primarily because of lower carrying capacity (MW per line) of electrical transmission lines compared to the* energy carrying capacity of the pipelines for gaseous and liquid fuels. The differences in the cost of transmission are important but often unrecognized and should be considered as a significant cost component in the analysis of various renewable energy production, distribution, and utilization scenarios."

Conditioning, transformation, network power frequency: The transmission and distribution losses discussed above are in addition to the conversion and conditioning losses that must also be specified. Wind turbines produce alternating current (AC) electricity that needs to be converted and "conditioned" or rectified. The alternating current from a wind turbine is not produced at a sufficiently stable voltage, frequency, or phase to insert directly into a grid. Typically, **an offshore substation rectifies and sums the current from individual wind turbines and transmits it to land. A wind farm's rectified electrical output must then be converted to the correct voltage, frequency, and phase before insertion into a grid.** Even after conditioning, wind and solar power in the grid does not have the same quality and is considered unclean or "dirty" by network specialists.

It is understandable that electricity generation from wind and solar is subject to significant fluctuations and can vary second by second. Accordingly, balancing must be done quickly via controlling power (German: "Regelleistung"). The more wind and PV plants are operated in the grid, the higher the demand for controlling power. It should be noted that a single undershoot of the maximum permissible grid frequency of 47,5 Hz in Germany inevitably leads to a blackout. From an underfrequency of 49 Hz, load is already reduced ("load shedding"), in other words "power is selectively switched off". Accordingly, controlling power is always provided by power plant types or consumers that are fully dispatchable and not dependent on fluctuating factors, such as coal, gas, or hydropower (Bleich 2022, netzfrequenz.info).

Due to the legal grid feed-in priority of solar and wind, more and more electricity enters the grid, which must first be conditioned for the grid with the help of rectifiers and inverters. In the process, uncontrollable "harmonics" can occur in addition to or above the standard frequency of 50 Hz (for Europe), which can lead to power losses due to heat generation and to other undesirable or even catastrophic side effects. On the other hand, uncontrollable amplifier effects can also occur due to resonances, which can lead to short circuits and fires. Destructive resonances can only be avoided if the proportion of "unclean" electricity in the grid is kept as small as possible. In Germany, on the other hand, the sources of unclean alternating current have been systematically increased for decades, resulting in "alternating current chaos" (Figure 13, Gaertner 2022).

Normally, electricity from individual wind turbines is rectified, summed up, and transmitted onshore from an offshore substation. *The rectified electrical power from a wind farm must then be converted to the correct voltage, frequency and phase before being fed into a grid.* As a reminder, for longer-distance transmission, the voltage is increased to limit resistance losses. Since the power, P, delivered via a transmission line is:

$$P = I * V = I^2 * R \qquad (2)$$

A higher voltage, V, lowers the current, I, and, in turn, reduces resistance, R, and therefore heating and power loss in the transmission cable. Rectification and subsequent inverter losses result in a round-trip loss of ~30% of the (on- and offshore) wind farm's raw electrical output. Onshore transmission losses alone are usually ~8-10%, depending on the distance from the wind turbines to the consumer, and additional rectification and inverter losses are unavoidable.

The transmission lines from deep water or remote land wind resources require high-voltage direct current (DC) lines to avoid AC reactance losses.

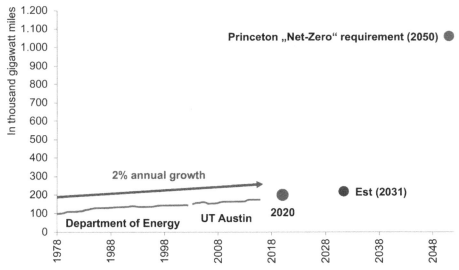

Figure 12: US transmission grid growth insufficient for "Net-Zero"
Note: US transmission grid in thousands of GW-miles has grown 2% p.a. since 1978.
Source: JP Morgan 2022, p12

2. Variable "renewable" energy and storage

The total conditioning plus transmission losses consist of 15% rectification, 15% inverter, and ~9% transmission loss (as noted for California by World Bank 2018), or ~34% of the raw wind farm output power, even over the relatively short distances in California.

As mentioned, variable "renewable" energy systems require vastly larger transmission and controlling power infrastructure. The size of the required transmission grid in Princeton University's "Net-Zero" pathway can be compared to the existing transmission grid in the US and is illustrated in Figure 12. There is little chance of reaching the required transmission grid size by 2050. If it were reached, the energy and materials required to build such grids would be tremendous and have neither been calculated nor modeled. The increased vulnerability of such vast transmission grids has not been considered either. As wind and solar installations reach higher penetration in the electricity grid, the number of disturbances increases. JP Morgan 2022 has summarized this information based on Department of Energy data in Figure 13. The situation is expected to be similar if not worse in Germany, the Western world's variable "renewable" energy champion.

Black start capability: Every power plant, every wind turbine, and every solar panel requires electricity to be able to "start up". The ability to start

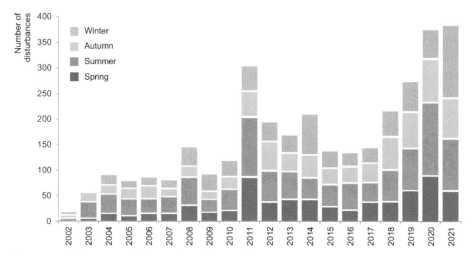

Figure 13: US reported electric disturbances by season
Source: JP Morgan 2022, p11, based on Department of Energy and JPM analysis

up without power from the grid, i.e., completely self-sufficient and thus as an island, is referred to as "black start capability". This capability is especially important in the event of brownouts or blackouts. The black start capability of larger power plants is provided by batteries that then start diesel generators, which are then used to ramp up power plants. In the case of wind and solar, batteries and diesel generators are also typically required. Hydroelectric plants and pumped hydro reservoirs usually require only a smaller amount of electricity to start up, for example, to open locks.

In Germany, during the first half of the 20th century, and also historically before that time, and in the former East Germany, every power plant had black start capability for good reason. In the 1980s, for economic reasons, this capability was not always planned for in newer power plants in the former West Germany. In Germany, the black start capability of power plants has thus been significantly reduced over the years. In the case of wind and solar, black start capability is rare to none-existent, although it is theoretically possible. As grid stability decreases with increasing wind and solar penetration, the risk of blackouts also increases.

When blackouts occur, the electricity grid must be slowly "ramped up" again. This is done in so-called grid islands, i.e., limited power consumers are connected to the grid together with a restarting power plant and the 50 Hz power frequency stability has to be built up. You remember that demand and supply must always be exactly equal for the grid to be stable. If the grid island runs constant at 50 Hz, it must be connected or merged to another grid island. This can only happen if both grids run completely synchronously. This synchronization of subnetworks can take days. If the synchronism between sub-grids to be merged is not given, generators and turbines can not only be damaged, but blow up. Large turbines are very sensitive when the frequency drops or rises, mechanical resonances can occur, which can cause irreparable damage.

2.4. Energy storage

To overcome the intermittency of wind and solar, energy storage is required. A future, efficient form of energy storage would also make the electrification of individual transport more sustainable. For electricity purposes, it is important to differentiate between three types of storage:

    a) Short-term storage, balancing in the second/minute range.
    b) Intermediate storage, balancing daily peak and low loads.

c) Long-term storage surpassing 2-12 weeks (Toke 2021). This is also often referred to as long-duration energy storage LDES (McKinsey 2021b).

The IEA refers to short-term storage as ramping flexibility and intermediate storage as *peak capacity/adequacy* (see Figure 33). *For Germany, Ruhnau and Quist 2022 estimate a long-duration energy storage requirement of 12 weeks or longer when analyzing the multi-year history of intermittent scarcity*, with the winter of 1996/97 constituting the low point in Germany's wind and solar availability. In 2021, there were lower than expected wind and solar resources in Northern Europe, resulting in a drop of wind- and solar-generated electricity across Europe and an 18% year-on-year increase in coal burn.

Long-duration energy storage (LDES) is the key challenge for the "green" energy transition. As part of their "Net-Zero" pathways, McKinsey 2021b estimate that, with large wind and solar penetration, around 10% of all electricity would need to be stored in some sort of LDES system by 2040, which clearly does not exist today. Before we speak more about storage, it is crucial for energy policymakers to understand and internalize that – as mentioned previously – although storage will make the integration of more wind and solar possible, *energy storage can never overcome the inherent physical challenges of wind and solar*, other than intermittency, namely (see Figure 8, p33):

- Point 2: Low energy density/space requirements.
- Point 3: Environmental damage to plant and animal life, and negative climatic impacts.
- Point 4: Low energy efficiencies.
- Point 6: Short lifetime.
- Point 8: High material input requirements.
- Point 9: Recycling challenges.
- Point 10: Resulting in low energy return on investment (eROI), accounting for all embodied energy of the total energy system.

Nevertheless, the search for a sustainable and affordable long-duration energy storage (LDES) solution at grid scale is at the forefront of energy research and remains important. LDES may come in different forms (McKinsey 2021b):

- ***Chemical***: storing energy through the creation of chemical bonds, for example, hydrogen.

- ***Electrochemical***: storing energy in batteries of different chemistries, for example, electrochemical flow batteries or air-metal batteries.
- ***Thermal***: storing energy through heating a solid or liquid medium and then using this heat energy later, also called Power-to-Heat-to-Power (PtHtP).
- ***Mechanical***: storing energy through kinetic or potential energy, for example, gravity based, pumped hydro, etc.

Despite billions of US dollars of already invested capital, and trillions of US dollars of committed future funding, energy storage challenges relating to a) energy densities, b) conversion efficiencies, c) lifetime, d) material input, e) recycling, and f) energy input (eROI) have not been overcome. In fact, due to the physics and chemistry of the proposed systems, there is relatively little upside to these already well-developed technologies. Please keep in mind that these energy storage challenges are similar and in addition to, but not the same as, the challenges for wind and solar discussed previously. Consulting firms and research institutions have published large reports on potential future storage solutions, but to date no truly sustainable and affordable grid-scale storage has been found.

Energy economics dictates that *any storage – which always adds complexity and requires energy transformation (see 2nd Law of Thermodynamics in the preface) – will always reduce the eROI and material efficiency of an energy system* (see Chapter 3.3). It should also be noted that any loss of energy due to conversion or storage directly results in low-value (high-entropy) heat and thus warms our biosphere, adding to measured temperature increases (Soon et al. 2015).

On molten salt and Power-to-Heat-to-Power (PtHtP): We are not able to cover all storage technologies but would like to say a few words about PtHtP and the molten salt storage technology that is commonly used. Bauer et al. 2021 summarized that concentrated solar power (CSP), also known as solar thermal electricity, is a commercial technology that produces heat by concentrating solar irradiation. CSP is expected to make up a significant share of "renewable" installed capacity and is preferred in sun-rich nations' "renewable" drives, such as in Pakistan, to replace fossil-fuel-based heat generating technologies. In fact, Lars Schernikau has personally visited solar and molten salt installations that make commercial sense in energy-deprived nations with high solar irradiance and without a fully functioning electricity grid.

Molten salt is most commonly and successfully coupled with concentrated solar power to allow for a few hours of storage. It must be noted that ***molten salt or PtHtP will not solve the long-duration energy storage problem as it lasts only hours or, in the best-case scenario, a couple of days***. High temperature thermal energy storage, however, appears to be one of the lowest cost medium-term energy storage solutions, according to Caraballo et al. 2021. Some consider coupling Power-to-Heat-to-Power with Power-to-Gas-to-Power technology, but in doing so the net round-trip efficiency plummets.

Molten salt technologies most commonly use "solar salt", a salt mixture of ~60 wt % sodium nitrate and ~40 wt % potassium nitrate. It comes with a specific cost of US$ ~1.300 per metric ton and a storage capacity or energy density of ~500 MJ/mt (Bauer et al. 2021 and Caraballo et al. 2021). The diverging round-trip efficiency estimates are interesting. Bauer et al. 2021, p543, estimate the round-trip efficiency of molten salt or PtHtP to reach ~85%, while MAN Energy Solutions 2021 assume ~35-40% round-trip efficiencies for their newest 100 MW molten salt sample project MAN MOSAS. This illustrates that ***researchers often overestimate actual real-life conversion or storage efficiencies***. Unfortunately, governments and institutions use such theoretical, unrealistic model calculations to model the large-scale "Net-Zero" employment of "renewable" energy systems worldwide, without sufficiently consulting the practical engineers who run today's energy systems.

On hydro and pumped storage as backup: Pumped storage means using electricity to pump water "up" into a reservoir and later "let it out" when needed to generate electricity. Of course, pumped storage requires access to mountainous regions, which are far from large cities such as New York, Washington, Chicago, Beijing, Moscow, London, Paris, Warsaw, or Berlin.

Norway's hydroelectric system is widely discussed as a backup for the German and European "renewable" energy systems. The undersea cables NordLink (NordLink, 2020), and earlier Nordned and Nordger, connect Norway's hydroelectric power to the Netherlands and Germany. Contrary to expectations, these connections can be viewed only as a minor backup for the "renewable" power in those two countries. Let us consider what potential Norway has for Europe.

Norway's hydroelectric facility is one of the world's best. Norway's hydroelectric energy storage capacity is 87 TWh, compared with a total annual energy output of 137 TWh, which is over 95% of the energy consumed by Norway (NVE Report 2021). These numbers show that precipitation must replenish around 50 TWh annually. The hydroelectric reserve itself varies by 60 TWh from year to year, so the net export capability of Norway can easily fall to zero.

Neither Germany nor Europe could possibly rely on Norway as its backup, even if the entire hydroelectric system were dedicated to German or European backup and nothing were provided to Norway itself. In the event of the success of "Net Zero 2050", the shortfall would apply to the EU's entire energy consumption and would be six times worse, corresponding to an only brief backup duration, assuming the power could be transmitted at all. In August 2022, Norway Minister for Energy Terje Aasland clarified: *"We are looking at how to limit exports in situations where reservoir filling becomes critically low. Then we must secure enough power for our national consumption"* (Montel 2022).

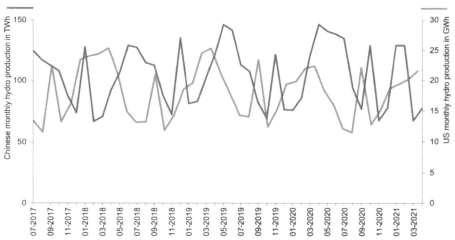

Figure 14: High correlation of output for hydroelectric energy in the USA and China Note: Statista (https://bit.ly/3AI86F3), Plazak (https://bit.ly/3wLiwCA) based on EIA. Source: Smith and Schernikau based on Statista and EIA

Figure 14 illustrates another profound limitation of hydroelectric energy as a backup. Like all wind and PV "renewable" energy, the backup from hydro is seasonal around the entire Northern Hemisphere (presumably also in the Southern Hemisphere) due to the seasonal variations in precipitation.

The figure illustrates the hydroelectric output of the USA compared to that of the Three Gorges Dam in China. It appears that the output varies coherently, with low and high output at the same times of year. This means that, even if power transmission over 20.000 km becomes feasible, China could not easily help the USA at times of low hydro backup, and vice-versa.

On lithium-ion batteries: For consumer purposes, we are already used to one-time-use and rechargeable batteries and know about batteries' tendency to self-discharge and lose capacity over time. It is now generally accepted that ***lithium-ion batteries (or any batteries used for EVs) are not the solution for long-duration energy storage*** (McKinsey 2021b) due to low energy density, self-discharge, raw material input, recycling issues, and so on. They are, however, sometimes considered for short-term storage. Figure 15 compares the energy density of modern lithium-ion batteries to the energy density of coal, as well as the material requirement for each. The authors realize that coal can only be "used" once while a battery can be charged many times, but the figure illustrates that the energy densities of fossil fuels are not in reach and that the raw material inputs for lithium-ion batteries are very large (see also the next Chapter 2.5 *Hydrogen and how it compares to hydrocarbons*).

We have prepared more detailed calculations showing what it would theoretically take to power 100% of Germany with solar PV in Spain, backed up solely by lithium-ion batteries. As a reminder, Germany consumes just over 15% of the power of the entire EU and is home to ~1% of the global population. The results, only for Germany's electricity needs and battery storage requirement, accounting for battery storage utilization factor of 1,7x and a conservative DC/AC conversion and transmission round-trip loss of 30%, are as follows (see Schernikau and Smith 2021 for calculations and sources):

- A moderate, yet insufficient, ***14-day storage backup for Germany during the winter would require ~45 TWh of battery storage***. Above, we compared that capacity with the entire Norwegian hydroelectric system. The output of ~900 Gigafactories, each producing 50 GWh p.a., would be required for construction of the batteries in one year.
    a) If the batteries lasted 20 years, the output of ~45 Gigafactories or 2,25 TWh would be required for annual battery replacement, in perpetuity (45 TWh/50 GWh/20 yrs).

b) For comparison, the replacement of batteries alone exceeds the 2021 global battery production capacity of 0,6 TWh by a factor of 4-5x.

- The raw materials required for 45 TWh of battery storage amount to 7-13 billion tons. Assuming a generous 20-year battery lifetime, *0,4-0,7 billion tons of raw materials would be required annually in perpetuity*. The materials required to build 2,25 TWh batteries annually in perpetuity for Germany include:

  a) ~6x global lithium production (~880 tons lithium per 1 GWh, 2020 production around 320.000 tons, ~70% from China),

  b) ~22x global graphite anodes production (~1.200 tons graphite anodes per 1 GWh, 2020 production around 210.000 tons, ~80% from China),

  c) ~2x global cobalt production (~100 tons cobalt per 1 GWh, 2020 production around 120.000 tons, ~80% from China), and

  d) ~8x global nickel sulfate production (~800 tons nickel sulfite per 1 GWh, 2020 production around 230.000 tons[6], ~60% from China).

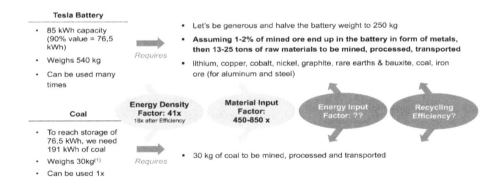

Figure 15: Illustrating energy densities – battery versus coal

(1) Assuming 5.500 kcal per kg for coal, 40% power plant efficiency to generate 76,5 kWh requires 191 kWh or around 30 kg of coal. Note: 1 kWh = 860 kcal = 0,086 kg oe = 3.600 kj; 1 kcal = 4,1868 kj.

Source: Schernikau Research and Analysis

---

6 Despite the total nickel market being large (over 2,3 million tons), only a fraction is geared to making nickel sulfate chemical for lithium-ion battery usage.

Currently, governments are focusing on "green" hydrogen H_2, which is discussed further in the next subchapter. Other potential storage solutions, such as but not limited to chemical-based or gravity-based storage, are also being researched but are not expected to be game changers at grid scale in the decades to come, though technological advances and positive surprises may change this. The case of the well-funded, multi-billion-US dollar valued Swiss EnergyVault exemplifies how celebrated "energy solutions" turn out to be, at best, of questionable environmental and economic benefit (CleanTechnica 2022). When investors realize such shortcomings, they quickly abandon the technology. The failure is then little discussed. Many startups, unfortunately, miss the opportunity to learn from such failures, stemming from disregarding net energy efficiencies across the entire value chain, and more investors' money is wasted.

Demand-side responses are not covered herein because demand-side response is not a storage solution but rather an optimization of energy utilization, which should be part of any responsible energy policy. Demand-side response may lead to partial storage (i.e., limiting charging and utilization of EVs during times of energy shortage) but has limits for a functioning society at large.

To store energy, fossil fuels or nuclear should not be used (i.e., in the form of blue or gray hydrogen) as they might just as well be used directly, together with or without CCUS technology. The latter might be much more environmentally and economically sustainable and would not require an additional storage medium, therefore reducing energy losses and increasing efficiencies. This little discussed reality becomes very important, as you will see.

Only excess – otherwise unused – intermittent "renewable" energy, such as wind and solar, should logically be used for feeding any future long-duration energy storage. Otherwise, it is always more economical to use the "green" energy directly. This logic also applies to low GHG-emitting geothermal, hydro, or nuclear energy. Since wind and solar suffer from the above-mentioned shortcomings, any long-duration energy storage will not truly solve the energy problem. However, a truly sustainable storage solution would optimize the use and distribution of available wind and solar energy, with all its shortcomings, and reduce the amount of fossil fuels combusted, as long as energy demand does not rise. We would like to remind you that mined energy fuels, such as uranium, oil, coal, and natural gas, are nothing but long-term energy storage systems that can only be used once.

We explicitly reiterate our support for continued basic research into environmentally and economically viable energy storage.

2.5. Hydrogen and how it compares to hydrocarbons

Hydrogen is one of the 10 most abundant elements on Earth. However, it is rarely found in isolation (HC Group 2021). To date, no one has seriously proposed deep drilling to obtain molecular hydrogen fuel directly. This means that hydrogen as we know it always needs to be produced, which requires energy. Therefore, its utilization as an energy carrier is expensive. Figure 18 illustrates that – generously calculated – around 60% of the input energy is lost when producing and "repowering" hydrogen using electrolysis and H_2-to-Power technology. Any transportation or storage of H_2 requires additional energy. *The total round-trip hydrogen storage system, including transportation and storage, may "cost" as much as 80% of the input energy.*

The hydrogen atom is a principal energy carrier in many chemical fuels because it is very reactive in accepting and releasing the energy in its chemical bonds with other atoms (Kiefer 2013). Furthermore, it is the lightest molecule, giving it a very high gravimetric energy density (joules per kilogram). On the other hand, H_2 has a very low volumetric energy density (joules per liter). A low volumetric energy density means that H_2 cannot be transported economically over large distances, nor can it be economically stored over the longer term.

Today, H_2 is considered or hoped to be the solution to long-duration energy storage and will attract trillions of US dollars in funding in the coming years. For decades, hydrogen has been produced and consumed primarily for industrial purposes. The size of today's traditional hydrogen economy depends on the definition of hydrogen used. Columbia 2022 summarizes the issue of insufficient statistics on H_2. Three categories of hydrogen are usually defined:

- Step 1: *pure hydrogen demand*, mostly used in oil refining and ammonia production (crucial for fertilizers/food production), estimated at around ~72 Mt as of 2020.
- Step 2: *hydrogen demand*, which includes pure hydrogen as well as around ~18 Mt of hydrogen mixed with other gases for the production of steel and methanol, total ~90 Mt.
- Step 3: *total hydrogen demand*, including pure hydrogen and a further ~45 Mt used in industry without prior separation from other gases (including the 18 Mt mentioned in the previous bullet point), estimated at ~120 Mt.

Today's hydrogen production plants can be divided into three main categories:
- *Captive production facilities*, where hydrogen is produced on-site for the plant's own consumption.
- *Merchant production facilities*, where hydrogen is sold externally.
- *Byproduct production*, where hydrogen is a byproduct of other processes (such as chlor-alkaline production).

Kiefer 2013 emphasizes that carbon is another lightweight element that is an excellent energy carrier and fuel component. Coal, which is primarily carbon, combusts in oxygen with an energetic output. **When it comes to hydrogen, carbon is a chemical miracle worker. Combined with hydrogen, carbon forms highly versatile and energetic hydrocarbon gaseous and liquid fuels.** Higher carbon ratios yield solids and lower ratios yield gases, all at typical ambient temperatures and pressures (see Appendix 1). Carbon also performs the trick of packing hydrogen atoms much more closely together. This explains why gasoline and octane contain over 60% more hydrogen atoms per m^3 than pure liquid hydrogen.

Because carbon adds its own significant energy to the mix, gasoline has ~3,5x higher volumetric energy density (joules per liter) than liquid hydrogen, and ~7x higher density than compressed H_2, as envisioned for H_2-powered cars (Figure 16). Only a few solids have a higher volumetric energy density than gasoline, which explains its wide use for transportation. The addition of carbon transforms hydrogen from a low-density, explosive gas that will only become liquid at around −250°C (20K) into an easily-handled room temperature hydrocarbon liquid, such as gasoline, diesel, or other petroleum products (CxHxOx). These hydrocarbons have more than triple the energy density of hydrogen alone (see Kiefer 2013 for more details).

Thus, hydrocarbon fuels are convenient to produce, store, transport, and consume, while H_2 is very difficult and energy-intensive to produce, handle, store, or transport for consumption and is highly explosive. Until a means to duplicate the utility of hydrocarbon fuels in a presently unknown H_2 or H-containing medium is found, hydrogen cannot solve the "renewable" energy storage problem, neither economically nor environmentally. Since the periodic table shows that discovering a new element to replace carbon is not probable, we must invent something else to replace carbon's role by finding ways to increase the volumetric density of H_2.

"If we didn't have carbon, we would have to invent it as the ideal tool for handling hydrogen" (Kiefer 2013, p117).

With today's technology, hydrogen's low volumetric energy density, the high energy cost of its production, its highly flammable characteristic, and its high transportation cost are a barrier to the widespread use of H_2. That is the reason why over 100 million tons of today's industrial H_2 are being utilized near their place of production and not being transported. Compressed H_2 storage requires heavy-duty storage cylinders lined with substances that H_2 does not permeate, so that they do not become brittle. Hydrogen embrittlement affects steel and alloy; it has been researched and is a serious issue for all hydrogen transportation and storage, as summarized by Professors Cairney, Hutchinson, Preuss, and Chen: *"The hydrogen embrittlement challenge is a highly complex materials and engineering problem"* (Cairney-RenewEconomy 2021).

In addition, more energy and raw materials are required to compress or liquefy and transport H_2. German energy economist Prof. Kemfert, who consults for the German government and supports Germany's "green" energy transition, points out that *"the production of H_2 requires 3-5x more energy than using renewable energy directly"*, stating that hydrogen is precious and should be considered as *"Champagne for Energy Systems"* (Kemfert 2021). Kemfert correctly points out that it costs 65 to 80% of input energy to produce, transport, store, and use hydrogen, in line with our analysis.

On the subject of transport, Bossel et al. 2009 concluded that, at 200 bar, a 40-ton truck delivers around 3,2 tons of methane but only 320 kg of H_2. This is due to H_2's low volumetric density and because of the weight of pressure vessels and safety armatures. Around 4,6 times more energy is required to move H_2 through a pipeline than is needed for the same natural gas energy transport. As mentioned, natural gas pipelines may suffer from H_2 transport because the ultra-light and highly volatile H_2 permeates steel pipes, making them brittle and increasing failure rates. Technological advances may overcome some of the above shortcomings through the development of materials which act as prolific H_2 "sponges", adsorbing H_2 and then releasing it for consumption (see further below).

ACWA in Saudi Arabia and many other companies plan to produce ammonia in combination with H_2 to ease the transportation burden of hydrogen (Air Products 2021). We have not evaluated this hybrid H_2-NH_3 concept but point out that any additional energy conversion again results in lost energy, thus warming the planet's biosphere and requiring additional processing equipment made from raw materials. Today, H_2 is almost entirely used as a chemical reagent (i.e., for fertilizer production, not as a fuel) near its production location. Over 98% of today's H_2 stems from fossil fuels without "carbon removal" technology (HC Group 2021).

2. VARIABLE "RENEWABLE" ENERGY AND STORAGE

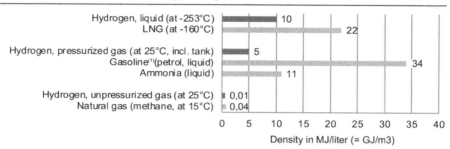

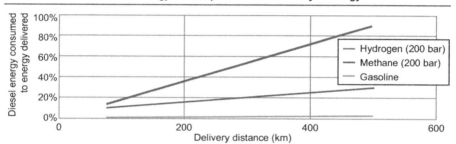

H₂ is 8x lighter than methane, to compress H₂ to 200 bar ~9% of its energy content is lost vs. ~2,5% when pressurizing methane.

H₂ is a synthetic energy carrier: High-grade energy is required to produce, compress, liquefy, transport, transfer or store H₂ (in most cases this energy could be distributed directly to the end user from wind or solar).

Figure 16: Volumetric energy densities of hydrogen, gasoline, and natural gas

(1) Diesel has slightly higher volumetric energy density than gasoline, at 38 MJ/l.

Note: Gravimetric heating value has little relevance for the hydrogen trade. The volume available for fuel tanks is always limited. Therefore, all practical assessments need to use volumetric rather than gravimetric energy density.

Note 2: Energy densities in Higher Heating Value, HHV, and rounded; gravimetric energy densities: H_2 liquid or unpressurized gas 140 MJ/kg, LNG at −160°C 55 MJ/kg, H_2 pressurized (inc. tank) 7 MJ/kg, gasoline and diesel 46 MJ/kg, ammonia (liquid) 19 MJ/kg, natural gas (methane, 15°C) 54 MJ/kg.

Note 3: For pressurized H_2, the high-pressure tanks weigh much more than the hydrogen they can hold. The hydrogen may be around ~6% of the total mass, giving just 7 MJ/kg total mass for the heating value; pure hydrogen gas holds 140 MJ/kg.

Sources: Bossel Eliasson 2006, Bossel 2009, Wikipedia on Energy Densities

The 3 GW electrolyzer in Namibia is an example of a hydrogen "megaproject" that is set to produce 300.000 tons of H_2 p.a. (H_2-View 2021). The US$ 9,4 billion project is scheduled to start operating in 2026. To put the project into perspective, it plans to produce less than 0,3% of today's global hydrogen output of just over 100 million tons. Another example is the US start-up Green Hydrogen International (GHI), which has announced a 60 GW or 2,5 million tons "renewable" H_2 project in a sparsely populated area of South Texas, to be powered by wind and solar. It plans to have the first 2 GW or ~80.000 t online (around 3% of GHI's total planned capacity) by 2026. With access to its own salt cavern, the company plans to produce clean rocket fuel for Elon Musk's SpaceX. *"Access to salt storage is critical to the scaling-up of green hydrogen production as it allows for maximum utilization of electrolyzers and serves as a buffer between variable wind and solar production and final delivery of green hydrogen to customers"* (Collins 2022a). Needless to say, **this "green" hydrogen for SpaceX would be used for additional, incremental energy demand for space travel, not for the replacement of existing fossil fuel infrastructure.**

S&P Platts 2021 keeps a global electrolyzer database and at the end of 2021 had accounted for around two million tons of global "low carbon" electrolyzer capacity. This output capacity is expected to rise to over 23 million tons by 2030. The IEA's "Net-Zero" pathway projects annual H_2 requirements of over 500 million tons by 2050 (IEA Net-Zero 2021), over 20 times higher than S&P projects for 2030. However, even 500 million tons may not be sufficient; most recently, IRENA 2022 and McKinsey 2022b ("achieved commitments" scenario) project ~600 million ton "clean" hydrogen production by 2050. **McKinsey projects that 28% of global electricity in 2050 will be used to produce "green" hydrogen.** All of this "green" H_2 will need to be transported from one region in the world to another (see Chapter 2.2 *Capacity factors*). Europe's climate chief Frans Timmerman admitted in May 2022 that "*Europe is never going to be capable of producing its own hydrogen in sufficient quantities*". In other words, it will have to be imported and transported over large distances, which, as we have explained above, costs significant energy (Collins 2022b).

Transporting H_2: In addition to the costs and losses associated with H_2 production, transport, and consumption, ***H_2 has a global warming potential, GWP_{20}, that is 33 times larger than that of CO_2*** (Derwent et al. 2006, Recharge News 2022b). Fugitive H_2 will grow quickly under the implementation of "Net-Zero", and the global warming potential will grow with the inevitable H_2 releases.

2. VARIABLE "RENEWABLE" ENERGY AND STORAGE

Figure 17: Arrival of the first liquid H_2 carrier in Australia in 2022
Source: Paul 2022

As previously mentioned, *only excess – otherwise unused – intermittent "renewable" energy, such as wind and solar, should logically be used to produce H_2 for the purpose of grid storage as otherwise it is always more economical to use the "green" energy directly.* Contrary to this logic, the world's first hydrogen tanker left Australia for Japan in the first quarter of 2022 carrying Australian hydrogen that was produced by reacting coal with oxygen and steam under high heat and pressure (Paul 2022, see Figure 17). The H_2 is trucked to a port site where it is cooled to −253°C to liquefy it for export. The US$ 360 million coal-to-hydrogen project is backed by Japan and Australia as a way to switch to "cleaner" energy and cut carbon emissions, although the question remains as to what happens with the carbon when coal is used to produce H_2.

We question the venture partners' environmental and economic logic. The partners are looking to produce up to 225.000 tons of hydrogen p.a. (0,2% of current global H_2 production, or 0,04% of "required" hydrogen as per IEA Net-Zero 2021). Shipping H_2 as fuel is crucial to the value chain for H_2, and this shipment was successful as a pilot project. We question the claim of a 100x improvement in H_2 daily loss (Recharge News 2022a) since vacuum vessels are not new technology. The vacuum-insulated tanks (dewars) used for shipping may have been pressurized to raise the boiling temperature and to slow the evaporation rate. Specifications are lacking of the operation. The ship did have a fire, unreported, while loading the liquified H_2 in Australia. The simple procedure for eliminating fugitive H_2 would be to flare the fugitive H_2 which only produces H_2O. Possibly the fire risk is too great, in that instance. While the vessel is underway, the H_2 can be used for propulsion either in a turbine combustion engine or a fuel cell.

It makes no environmental or economic sense to use coal or any dispatchable energy resource to produce H_2; it wastes energy that could be used elsewhere. Using such so-called blue or gray hydrogen for the purposes of power reduces energy efficiency and thus increases the negative environmental burden (Figure 18). In fact, such projects starve the world of much-needed energy at the time of a major global energy crisis (see Chapter 4.2, Footnote 10).

Significant research has been carried out and progress made in recent years in relation to what are known as "hydrogen sponges" (Morris et al. 2019 and NU 2020). Some candidates appear to reach 8% by weight of H_2. The materials used are relatively inexpensive and abundant, such as transition metals and carbon lattices as a scaffold for the metals. In the not-too-distant future, hydrogen sponges may become an appropriate medium for storing H_2 in a denser manner, presenting a potentially viable alternative to lithium-ion battery storage. The use of H_2 clathrates is another potential method for H_2 storage and transport that may become more effective and cheaper. To date, this has not received the same level of attention but may offer advantages (Gupta et al. 2021). This method can use methane, water, and other substances offering lower energetics and pressure, higher temperature stability, and less dangerous matrix storage compared to carrying hydrogen as ammonia, for example.

Around three quarters of energy is lost in producing, transporting, storing, and using hydrogen, and this lost energy will end up in high-entropy heat that warms our biosphere and always reduces the energy efficiency of the entire energy system (Figure 18). This high-entropy waste heat and energy inefficiency when producing "green" hydrogen arises primarily from the intermittent, low-eROI, and thus energy-inefficient wind and solar sources. At the same time, we explicitly support waste-to-energy and possibly waste-to-hydrogen technologies and investment, where they make sense. Human waste is probably one of the largest challenges that our society faces. Scavenging energy from and recycling human waste becomes more important as population levels and living standards rise. However, waste is not always composed of combustibles and includes many toxic pollutants. The heat content is often small compared with hydrocarbons and biomass fuels, and recycling requires energy.

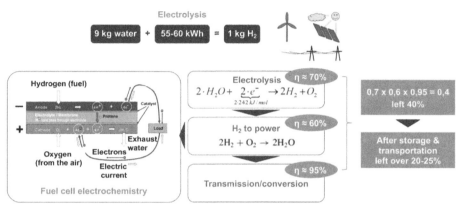

Figure 18: Hydrogen electrolysis and methanization (illustration)

Note: Methanization has an efficiency of 30-40% and combustion of methane about 45-50%. Thus, in the case of methanization, total efficiency goes down to around 12%.

Source: Schernikau Illustration based on Prof. Holger Watter (Watter 2021)

2.6. Material input and embodied energy

The material efficiency of energy systems is a key environmental consideration. It must be humanity's goal to become more material efficient in all our endeavors. Everything we produce and everything we consume ultimately originates from "raw" materials that are mined or grown. **Grown minerals** are considered "renewable" because they can grow back and are not finite in the same way as **mined minerals**, and there is truth in this. However, the rate at which grown materials can be consumed is limited by the rate at which they can be replenished. The frequent suggestion that biomass can increase to supply humanities' material demands is inconsistent with the total global production of organic carbon. That number is fixed by the ecology and the solar input.

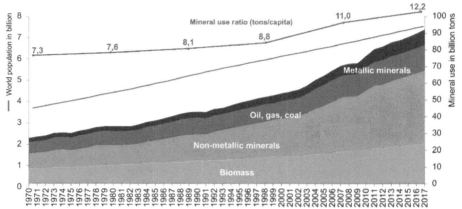

Figure 19: Global grown and mined material extractions from 1970-2017
Source: Schernikau Illustration based on WU Vienna 2020, UN 2019

Figure 19 illustrates that humanity today requires almost 100 billion tons of grown and mined minerals for its existence. The material consumption per capita has continuously increased, which so far has been largely driven by the reduction of poverty. What can energy contribute to improving material efficiency while further eradicating poverty? ***Classic fossil energy raw materials that contribute to over 80% of global primary energy make up ~15% of all minerals extracted.***

Figure 19 includes biomass because planting, harvesting, processing, and transporting **grown minerals** – which many consider "renewable" – also requires energy, machinery, labor, and space, which again negatively affects the environment (see Chapter 3.2 for a discussion on net energy efficiency or energy returns – eROI). The biomass we grow is mostly used for feeding livestock and the global population, but it is increasingly also used for energy generation. Germany generates about 7-8% of its electricity from biomass.

Mined minerals or raw materials are extracted from the ground, usually transported, and then processed before consumption. Coal is one of the few raw materials that does not always require processing before consumption and is often combusted directly. However, it must be noted that oil, coal, and gas are at the very core of our modern life systems, providing over 3/4 of all the input energy required today to produce consumables and services, including those used for the "green" energy transition itself. They are required to mine, process, transport, and upgrade the remaining ~85% of minerals and raw materials illustrated in Figure 19, namely

2. Variable "renewable" energy and storage

biomass, non-metallic minerals, and metallic minerals, which then form the material basis for those consumables.

When we speak about material input, we often consider steel, copper, cement, glass, aluminum, or silicon as base products. However, as mentioned above, these products are first produced from mined raw materials such as iron ore, coal, limestone, silicon, bauxite, quartz stone, etc. The processing of these raw materials into "base products" takes energy, which is called embodied (or sometimes embedded) energy. The acclaimed book *Sustainable Materials without the Hot Air* by Cambridge researchers Allwood and Cullen 2015 summarizes the concept of embodied energy. Figure 20 illustrates the embodied energy of selected industrial materials (perhaps more appropriately referred to as "base products") and shows that aluminum requires 5x more energy per ton to be produced than steel. It must be our goal to find novel technologies to reduce the energy input required to produce these materials, rather than making the energy required to produce them even more material intensive, as illustrated in Figure 19. Please note that Figure 20 does not show the material input required to produce these materials, only the energy it takes.

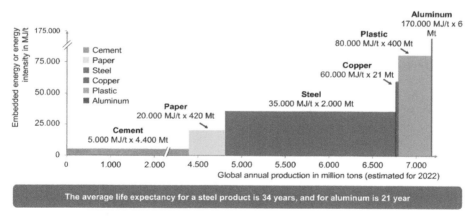

Figure 20: Embodied energy of selected construction and consumption materials

Note: All numbers are approximate.

Source: Schernikau based on Sustainable Materials without the Hot Air by Cambridge researchers Allwood and Cullen 2015; annual production for 2022 based on worldsteel.org, statista.com, international-aluminium.org; copper-embodied energy estimated from Rankin 2012 and Pitt and Wadsworth 1981

As an example: Steel takes around 35.000 MJ per ton to be produced. Steel is an alloy of iron and carbon; it can also contain small quantities of silicon, phosphorus, sulfur, and oxygen. Each ton of steel requires ~1,6 tons of iron ore, ~800 tons of coal, and products such as limestone and various additives; let us approximate the total to ~2,5 tons of input raw materials for each ton of steel. Therefore, the ~2 billion tons of steel to be produced in 2022 will require approximately 5 billion tons of raw materials to be extracted, transported, and processed. Thus, steel production appears to consume a little more than 5% of all global raw materials[7].

For chemical reduction purposes, when certain raw materials are involved (e.g., iron ore to iron, silicon oxide to silicon), a metallurgical coal product is required to bind the oxygen with carbon and release it in the form of carbon dioxide or carbon monoxide.

Simplified iron ore to iron: $2 Fe_2O_3 + 3C => 4 Fe + 3 CO_2$ (3)
(in a multi-step process)

Simplified sand to silicon: $SiO_2 + C => Si + CO_2$ (4)

For example, silicon – one of the key building blocks for solar panels and also computer chips – is essentially produced from silica (quartz stone), wood chips, and coal. Silicon is the second most abundant element in the Earth's crust, but so far only high-purity silica (quartz stone) is commercially viable. Metallurgical-grade silicon (MG-Si, around 98% purity) is manufactured at a process temperature of more than 2.000°C in electric arc furnaces, which also require coal for reduction and energy. Both solar-grade silicon (SoG-Si, 99,9999% purity) and electronic-grade silicon (EG-Si, 99,9999999% purity) are then produced out of metallurgical-grade silicon in a refining process ("Siemens" or other processes), which again requires large amounts of energy and various chemicals extracted from raw materials (pv-education 2020). Solar panel production requires various manufacturing stages, which include the production of polysilicon, ingots, wafers, PV cells, and finally the PV modules.

The surface mining of raw materials requires large amounts of overburden to be moved aside before the sought ore body can be reached. **It is not unusual to have an average "overburden ratio" of 10:1, which means every ton of raw mineral requires 10 tons of earth to be "mined" and removed.** If one now assumes a generous 2% average metal content in the

---

7   All numbers are approximate, rounded, and for illustrative purposes.

mined ore, e.g., for copper, and a 10:1 average overburden ratio, one then needs to mine 500 tons of earth for each ton of copper base mineral. For the copper to be inside your electric vehicle or wind turbine, transportation across the globe and various processing steps using low-cost energy in foreign nations is required.

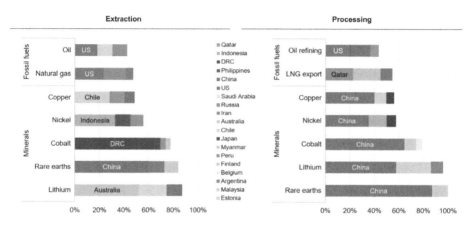

Figure 21: Share of top three producing countries in production of selected minerals and fossil fuels
Note: 2019 data.
Source: The Role of Critical Minerals in Clean Energy Transitions, IEA Minerals 2021, p13

Today, China is the base minerals processing hub of the world (Figure 21). China controls not only almost 80% of solar panel production but also uses partially forced labor (Murphy and Nyrola 2021) and mostly low-cost electricity produced from coal for processing and production. IEA Solar 2022 confirms: *"China now leads the [solar PV] market once dominated by Europe, the United States and Japan"*, raising significant geopolitical concerns as *"China significantly dominates every single solar PV supply chain segment"*. Today, all top 10 solar PV manufacturing firms that account for almost 50% of global market share are in China.

With this information, it may become apparent to the reader that "renewable" forms of energy generation are hardly truly "renewable". The material demands of wind, solar, and hydro power stem from their low energy density, large space requirement, and other physical shortcomings driven by the laws of energy, as discussed in earlier and later chapters (Figure 8 and

Figure 32); this results in an "incredible" material inefficiency. The US Department of Energy (DOE) has estimated the base-material input per 1 TW generation capacity in 2015 (DOE 2015), and this is summarized in Figure 22. It can be seen that the production of the generation equipment for "renewables" requires a multiple of refined materials. These refined materials then require a multiple in terms of raw material input and overburden removal as discussed earlier. Figure 22 does not account for combusted fuels, nor for the differing lifespan of the electricity production equipment. Kalt et al. 2022 also confirm the material inefficiencies for "renewables": *"We find robust evidence that scenarios in line with the 1,5°C target are associated with significantly higher material requirements than scenarios exceeding a global temperature rise of 2°C."*

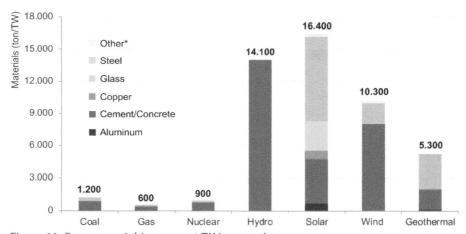

Figure 22: Base-material input per 1 TW generation

Note: Other includes iron, lead, plastic, and silicon; Schernikau assumes this is based on average US capacity factors.

Source: Adapted from DOE 2015, Table 10.4, p390

The IEA (IEA Minerals 2021) and many other renowned consulting firms and agencies (BCG, McKinsey, KU Leuven, IEEJ, EIA, S&P, etc.) have brought out more recent publications than DOE 2015 on the mineral and raw material needs of "renewables". Figure 23 compares not only power generation technologies but also electric vehicles and conventional cars in terms of their selected refined material requirements based on more recent information. Because many of the materials to build "renewables" are processed in China

(Figure 21), it appears that Western nations are shifting energy dependence from fossil fuel producers such as Russia and the Middle East toward mineral processing champion China and other countries with low-cost access to energy to provide the material basis for the "green energy transition". This shift of power will have further geopolitical consequences.

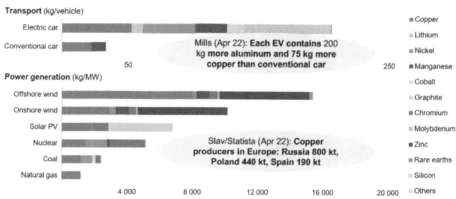

Figure 23: Comparing mineral needs for "renewable" technologies
Source: The Role of Critical Minerals in Clean Energy Transitions, IEA Minerals 2021, p6

Differentiating between energy-"producing" technologies (wind, solar, hydro, geothermal) and energy-consuming technologies (electric vehicles, heat pumps, DRI, etc.), it appears that copper, silicon, zinc, chromium, nickel, and rare earths are the key minerals required for the "energy transition" toward "renewable" forms of electricity production. The material and energy requirements just for building electricity networks are, however, often overlooked (Figure 24). Concerning copper, S&P 2022 summarize in their newest report on the global copper market: *"Copper – the 'metal of electrification' – is essential to all energy transition plans. But the potential supply-demand gap is expected to be very large as the transition proceeds".* **Substitution and recycling will not be enough to meet the demands of electric vehicles (EVs), power infrastructure, and renewable generation."**

Another important point is that materials projections are critically dependent on calculating realistic areas for wind farms and PV parks that meet the actual demand plus backup charging. Commonly, materials demands are grossly underestimated since the essential space and installed capacity requirements are misjudged, as our papers clearly demonstrate (Schernikau and Smith 2021 on solar, Smith and Schernikau 2022 on wind). Some studies suggest similar space requirements to those we have calculated, but most studies first estimate the land or sea area available and then assume the energy collection facility can somehow be fitted into the permitted space. Physics denies that approach.

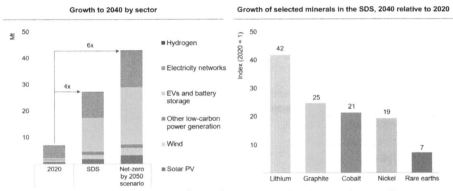

Figure 24: Mineral demand for "clean" energy technologies drastically increases depending on the scenario

Note: SDS = IEA's Sustained Development Scenarios; Mt = million tons. Includes all minerals in the scope of the IEA report (right-hand graph plus copper) but does not include steel, aluminum, cement, or any of the energy raw materials required to produce the metals.

Source: The Role of Critical Minerals in Clean Energy Transitions, IEA Minerals 2021, p9

Vaclav Smil is one of Bill Gates' favorite authors and his advisor on energy. In 2022, he published the widely recommended book *How the World Really Works: The Science Behind How We Got Here and Where We're Going* (Smil 2022). Smil clarifies that humankind uses 17% of the world's primary energy supply to make just four materials – ammonia (for fertilizer), steel, cement, and plastic (see also Figure 20). These substances, Smil explains, are *"pillars of modern civilization"*, crucial to feeding, housing, transporting, and – through medical devices or hospital construction – healing billions of people. Not only are there no readily available substitutes for these materials, but there are also no practical low-carbon ways to produce enough of them to meet current demand. And the world is actually going to need to expand its production as Africa and Asia modernize (Lane 2022).

The dramatic expansion of variable "renewable" energy capacity, quadrupling today's entire installed electricity production capacity (see Figure 34 on page 102) ***will result*** not only in energy starvation as explained later but also *in a raw materials crisis or, to put it more positively, a "new mining super cycle", which started in 2021* and which we are currently living through. Prices of raw materials and components of all kinds started to rise in 2021 and continued to reach new highs in 2022. Polysilicon prices are just one example; they more than tripled from US$ 10/kg in January 2021 to US$ 34/kg in April 2022 (IEA Solar 2022, p66). Fossil fuel and ore product prices also dramatically increased.

Raw material producers will earn large amounts of money when demand increases cannot be met by the investment-starved fossil fuel supplies, which are required for mining, processing, transportation, and equipment manufacturing. These price increases are to be borne by the final consumer, either in the form of higher taxes or directly though product price increases. These price increases will lead to inflationary pressures, slower growth, and specifically hurt the less well-off population in the West and developing nations as a whole.

The sometimes proposed idea to increase mining and energy companies' taxes for doing their job and benefiting from misguided energy policies, in order to then pour this money back into the "green" energy transition, seems illogical and would make the situation worse and worse. It is illogical because the high raw material and energy prices are a direct result of the prescribed "transition" and the lack of investment in conventional energy, as today's energy policies disincentivize long-term projects. Mining and energy projects require large investments that demand multiple decades of planned revenue streams. The only logical solution would be investment in, not divestment from, energy raw material production and conventional energy generation technology.

Material efficiency, and with it sufficient investment in energy raw materials such as oil, coal, and gas to produce these materials as energy efficiently as possible, is an important environmental metric to be considered when prioritizing energy systems. *The transition from conventional energy to "renewables" requires significantly more minerals and materials per unit of produced energy* than staying with current systems (see also Kleijn et al. 2012). This "green" transition requires new, or different, minerals that need to be mined, processed, and transported using production capacity that currently does not exist, as well as energy that is currently not available or can only be provided with conventional energy generation.

- IEA Minerals 2021, Executive Summary, warned: *"The shift to a clean energy system, is set to drive a huge increase in the requirements for these minerals... A rapid rise in demand for critical minerals – in most cases well above anything seen previously – poses huge questions about the availability and reliability of supply."*

2.7. Primary energy and heat pumps

Primary energy

Discussions about the use of the primary energy metric have emerged as "renewables" such as wind and solar make up a larger share of electricity but a lower share of primary energy, as mentioned in earlier chapters. *Some economists consider primary energy outdated and misleading because they, in our view mistakenly, consider that "primary energy" from solar and wind can be converted to usable electricity with little energy loss.* For this important discussion, we clarify that primary energy used for electricity generation (around 40% of PE) translates to ~28.000 TWh of usable electricity. This difference results from the conversion efficiency of conventional fuels to electricity in thermal power plants. In the same way, primary energy can only lead to usable energy for heating, transportation, or industrial purposes through lossy processes, as per the second 2nd Law of Thermodynamics. Remember that all primary energy sources always require some of their energy to be processed into finished fuels or electricity.

Often the argument against primary energy is that, in the future, wind and solar will produce electricity directly "without conversion losses" or further energy input, and heat pumps and EVs will use electricity at much higher efficiencies than conventional heaters and internal combustion engine cars. Therefore, so the argument continues, wind and solar electricity's generating share of primary energy is artificially low, and comparing investment and output using the primary energy metric appears misleading. The analogous argument is that an electric vehicle uses around one third the energy of a standard combustion vehicle but provides the same service. All these arguments are – in our view – either irrelevant, incorrect, or at least misleading, as we shall illustrate. To confuse the matter further, there are different ways of calculating primary energy: the partial substitution method and the physical energy content method (see OurWorldInData 2020 and IEA Statistics 2019).

Our reasons for not viewing the primary energy metric as outdated or misleading are grounded in the principles of thermodynamics. *Primary energy's importance arises from a different perspective when considering the overall energy efficiencies (eROI) of variable "renewable" energy sources such as wind and solar. It arises from the energy requirement across the entire value chain.* Despite its shortcomings, we consider primary energy to be the best way of estimating the total energy requirement for human existence. Positive effects of electrification certainly exist, but electricity produced by "renewables" still needs to be created in accordance with the laws of thermodynamics. "Renewable" energy, be it wind, solar, hydro-, or geothermal, is remarkably capable when efficiently transformed into electrical energy for powering our lives (including both for transportation and heating), but it is not magic. A clear account of the advantages as well as the limitations of the "renewable" energy used to create electricity is essential in order to make correct and sustainable choices that will allow us to meet our civilization's future energy demands.

It is necessary to acknowledge that electricity is a secondary (or even tertiary) form of energy. Also, conventional power plants often generate additional, valuable, non-electrical energy in the form of heat, which is then used for housing and industrial purposes, as a byproduct of electricity generation. Let us examine the assertion that electricity is produced from solar and wind without significant conversion/conditioning losses, or without significant energy costs, which turns out to be incorrect. The following numbers/calculations for wind come from Smith and Schernikau 2022, which includes all references and sources:

1. The maximum efficiency for wind turbines (WTs) is the Betz Limit of 59,2%, but none reach that performance level. Real-world WT efficiency is well below the Betz Limit: High-speed WTs are designed to achieve their "peak efficiency" at an optimum wind speed of around 14 m/sec, which yields 35-45% in terms of energy extraction to the turbine. Efficiency declines at wind speeds above and below optimum, and wind speeds rarely reach this optimal level.

    a) Wind turbine components convert the rotation of the WT rotor to electricity via gearboxes, bearings, generators, etc., into electrical output of the WT, further reducing the efficiency.

    b) The electrical output at the wind turbine location, then, lies between 10-30% of the raw wind power available, but only for the design wind speed. A WT's "installed power" of 10 MW (electricity at the wind farm) then requires 30-100 MW of raw wind power. The loss is thus between 70-90% of the wind's raw power. This 70-90% loss of the wind's raw power directly translates into the energy, material, and space inefficiency of wind (and solar).

    c) One can now argue that wind and sun are free, which is correct, but in a way coal, gas, and oil are also "free", in that they are provided by nature and just need to be "dug" out of the ground. Of course, fossil fuels are finite (over hundreds or thousands of years), while wind and solar appear not to be.

2. Building 3-10 times as many wind turbines in wind farms will not increase the electrical power correspondingly. The withdrawal rate of power from the wind is limited to the downward transfer of power from the upper atmosphere – around 1-2 MW/km^2. It is also limited by the wind speed loss at the first row of wind turbines and wake losses incurred, which demand a separation of WTs by 15 rotor diameters.

    a) The withdrawn energy is strictly limited by atmospheric potential and kinetic energy, which is a direct consequence of global solar heating of the atmosphere.

    b) The result is an upper limit on wind turbine power density. The construction of large wind farms scheduled to reach densities of up to 30 MW/km^2 results in an increasingly poor capacity factor at the site, reducing relative electricity output. Wind farm output has already demonstrated losses of almost ~30% at measured sites due to this effect when wind turbines are installed at too high density. This trend will worsen with ever-larger wind farms (Smith and Schernikau 2022).

3. Additional loss of "renewable" power has been demonstrated in California and Europe. Unlike fossil fuel or nuclear thermal plants, "renewable" energy plants are sited hundreds to thousands of kilometers from the customer, often along a seashore. These large distances require the construction of vast and vulnerable transmission infrastructures (see Figure 12). This vulnerability was demonstrated by the destruction of California transmission lines by forest fires. The round-trip conversion loss of the A/C produced by wind turbines to D/C and back is typically ~30% or more.

    a) Data for California's photovoltaic (PV) parks show a similar 25% loss due to transmission plus conversion to A/C; such losses are not encountered with base power plants since phase, frequency, and voltage can be controlled at the plant and inserted directly into the grid. Undersea cable losses are similar, plus they are vulnerable to turbidity currents, scour, and other displacements and breakages, resulting in long costly outages (see also Chapter 2.3)

4. Essentially all wind and solar installations require 100% backup, either in the form of batteries, "green" hydrogen, pumped hydro, or –

in most cases – an idle gas- or coal-fired power plant on standby. The energy spent on constructing, operating, and recycling these backup/storage systems needs to be included and can only be truly captured by the total primary energy "spent" on building our entire energy infrastructure, not by the measured electricity output from wind and solar alone. Keep in mind that these backup systems become more energy inefficient the less they are used as wind and solar penetration increases (see next chapter). To understand the impact of underutilized backup systems, we recommend a 2-minute German video on a backup gas-fired power plant in Bavaria, Germany, that runs for less than 10 days a year (NDR3 2020).

5. "Renewable" power plants convert a small fraction of the input energy into electricity at the consumer's home due to this series of losses. The losses are fundamental and due to thermodynamics, friction, resistance losses over long-distance transmission, repeated conversions, vulnerability to breakage/destruction, plus the much greater requirement for backup and storage in order to match power demand.

   a) A wind turbine, "operating at a 50% capacity factor", counts only the time at saturation output compared to the time below saturation, in effect neglecting all other losses. This neglect contributes to the high cost of VRE (see Chapter 3. *Cost of electricity and eROI*), contrary to what a marginal cost measure such as LCOE may indicate.

The above illustrates that the benefits of variable "renewable" energy must be weighed against the energy and cost inefficiencies of intermittent "renewables" because of backup and storage requirements, low energy efficiencies, conversion losses, material inefficiencies, shorter lifetime, higher recycling challenges, and room costs. **Only primary energy can capture the input energy required for any energy system including the energy losses from all the above-mentioned inefficiencies.** Energy costs and energy returns (eROI) are discussed in more detail in the next chapter.

- A recent statement by Thyssen Steel Chairman Osburg exemplifies the importance of net energy efficiency: *"Going climate neutral will increase energy demand 10x from 4,5 TWh to 45 TWh"* just for Europe's largest steel plant, in Duisburg (Dierig 2022, see Chapter 4.1).

In summary, it is incorrect to assume that wind and solar are more net primary energy efficient than conventional power plants. Wind and solar do not produce heat as a byproduct, and they require more material input, more space, and have lower net energy efficiency. Their adoption will require more, not less, primary energy for the same output of electricity when employed at large scale and as envisioned by "Net-Zero" pathways.

Heat pumps (we do not discuss cooling)

Since heat pumps are often singled out as a good example of the efficient use of electricity, let us examine the physics of the ideal heat pump. Heat pumps are envisioned as replacements for fossil fuels (oil, coal, gas), heating our homes by utilizing "renewable" electricity mostly produced from wind and solar to run the heat pump. Figure 25 shows how a heat pump uses work, w, to move heat Q from a cold to a warm place (GSU 2021). *Heat pumps have a well-deserved place in energy systems because moving heat from place to place takes less work (or energy) than cooling or heating directly* using fossil fuels. In fact, depending on the surrounding conditions, moving heat from one place to another may only take one third of the energy of cooling or heating directly. This is where many analysts stop and say: *"See, I told you so, it is more efficient."* But let us examine heat pumps in more detail.

The first and most important point is that a heat pump cannot function without electricity. Electricity is required to move the heat Q from one place to another, as described above and in Figure 25. This electricity needs to be securely available day and night, especially when it is cold, which is when heat pumps are most needed. This is crucial since heating our houses has been provided (a) from thermal power plants, which produce heat as a byproduct of electricity or (b) from combusting gas or oil (historically more coal) directly in or near the home. Heat or steam for industrial purposes originates almost entirely from gas and coal combustion. Back to home use:

- A *heat pump with a ground source* works well unless the ground freezes due to the extraction of too much heat from too little ground volume. Then, the heat pump or heat exchanger becomes very inefficient, leading to a loss of heating capability so that the ability to heat the home interior drops drastically (Figure 26). In colder conditions, the heat pump advantage is lost, and it barely functions. That is a problem in countries where summers are cool and winters are cold. The heat taken from the ground is never replaced by diffusion of heat into the cold volume.

2. Variable "renewable" energy and storage

- In cities, there is too little space for in-ground heat pumps, especially in large buildings or apartments; thus, air exchange is used. Heating the interior of an apartment building in a city may be accomplished by heat being withdrawn from the air. The **air exchange heat pump** drops the temperature of the outside air as heat is extracted into a building (the opposite of a freezer). In cold weather, when heat is needed most, moisture in the air freezes onto the heat exchanger.
- Air source heat pumps lose efficiency at temperatures below ~7°C and normally do not function below −4°C, so auxiliary heating, usually energy inefficient resistance heating, is then required. The same applies if you are heating the interior of a home and the heat is withdrawn from the air: The performance efficiency of the heat pump (Coefficient of Performance CP, Figure 25) can drop to near unity during cold spells (a fact not advertised by the heat pump salesman), and the heat exchanger will freeze up as moisture is deposited from the air.
- This heat exchange is also what refrigerators do, but the outcome is the opposite: Heat is expelled into the air. The only difference between a heat pump and an air conditioner is a value which controls the direction in which heat is pumped. Naturally, a heat pump typically lasts a shorter time than an air conditioner since it is working in both summer and winter.

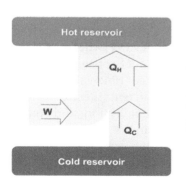
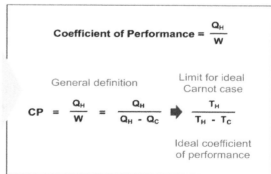

Figure 25: Physics of a heat pump
Note: W = Work, Q = Heat moved, T = Temperature, H stands for Hot, C stands for Cold.
Source: Georgia State University (GSU 2021)

The *Coefficient of Performance* (CP) is equal to the ratio of heat moved from outside to inside your home compared to the work (or energy) required to move it. The maximum occurs in the ideal Carnot cycle case, which is not practically obtainable, but as T_H and T_C approach each other, CP becomes very high. An electric heat pump, at least in the southern United States, allows a CP of at least 3. This means that three units of heat can be pumped into a house with the expenditure of just one unit of high-quality electric energy, which would appear to be the desired result. The exact CP, of course, depends on the difference between T_H and T_C and the specific design parameters of the heat pump. When the outside temperature falls below 0°C, the heat pump drops in efficiency, as noted above. Then, auxiliary resistance heating is required. ***This means that the heat pump performs well when little heating is required, but it starts to underperform when more heating is required*** (Figure 26). This is the situation in every heat-pump-equipped home and is especially relevant in a climate where it snows.

The assertion of 100% efficiency in terms of the use of electricity to produce heat is a common claim by electric utility companies. It is misleading because up to ~2,5-3 units of primary fuel, such as coal or gas, have to be burned to deliver 1 unit of electric energy to the house; this is due to the Carnot Cycle efficiency in electricity generation. Thus, 100% efficient use at your house is around 30-40% efficient in terms of the use of the primary fuel. If you were to use 100% wind or solar, this conversion loss from primary energy to electricity appears to become smaller. However, we discussed the primary energy argument in the previous section of this chapter. It is impossible to achieve 100% efficiency because of wind and solar's intermittency and unpredictability, and again you have to account for backup or storage, conversion and transmission, and the other inefficiencies detailed in Chapter 3 and above. The primary fuel consumption tells the real story, and the use of electricity from variable "renewable" sources is no different. VRE is on average between 10-30% efficient (see 2.2 *Capacity factors*) as it also loses energy due to transmission, conversion, and conditioning. In the real world, all energy sources have similar costs due to the laws of physics.

If you heat your home with natural gas, you are using the primary fuel directly at your house. This is preferable to using electric resistance heating, especially on cold days, when it would be a waste of the delivered high-quality electric energy. The often subsidized price charged to the consumer for electric heat pumps may be initially lower than the price charged for natural gas heating, but price does not equal cost to the system (see Chapter 3) and is skewed by subsidies for electricity from "renewable" resources compared with natural gas. Over the last 25 years, natural gas and electric heat pump heating have stayed comparable in cost, at least in North America, where natural gas prices have been stabilized by the abundance created by shale fracking.

Another consideration was illustrated in February 2021, when the electricity network collapsed in Texas (ERCOT 2021). Heat pumps were rendered inoperable and no EVs could be charged. Even worse, the pumps supplying natural gas were also operated using subsidized electricity from wind farms. The result was that no heat at all was available, neither from "renewable electricity" nor natural gas. The human cost has still not been accounted for.

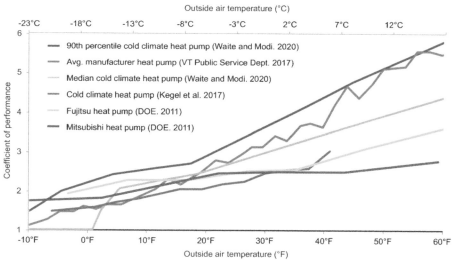

Figure 26: Heat pump performance reduces as it gets colder
Source: JP Morgan 2022, p23, based on JPMAM 2021

Thus, it appears that EVs or electric heat pumps use less primary energy than an ICE (internal combustion engine) vehicle or a gas boiler. When fairly considering the full life cycle and entire energy system, that picture looks very different. However, it is complicated to quantify uniquely. Full life-cycle cost to the entire energy system (see Chapter 3) must include the material and energy required for raw material production, processing, transportation, construction, operation, and the recycling of the actual equipment as well as the under-utilized backup systems. More independent research is required to establish realistic energy, monetary, and environmental costs.

Only primary energy can determine the true and total energy input for producing, using, and recycling our (energy) infrastructure. It must be humanity's aim to increase energy and material efficiencies, thus increasing eROI and reducing material input.

3. Cost of electricity and eROI

Lars Schernikau has completed a total of over 70 interviews in Europe, Africa, Asia, and North America during the past three years. Discussions have taken place at various ministries, governmental economic organizations, universities, and industrial conglomerates. The overarching theme from these interviews was a lack of understanding of the true full cost of electricity and the continued misuse of the marginal cost measure LCOE to compare costs of variable "renewables" with conventional sources of power. In all interviews, the overarching desire – especially in developing nations – was to support a sustainable yet economically viable energy policy to transition away from fossil fuels over time. The costs and downsides associated with this transition – limited by today's technologies – were rarely understood or researched.

Lars has also contacted energy think tanks such as the IEA, IEEJ, IMF, and ACE (ASEAN Center for Energy), as well as leading strategy consulting firms, and discussed some of the above topics with them. The conclusions herein are a result of these interactions, as well as further research. The political aspect inherent in the work of all of the above-mentioned organizations was pointed out to us but is not discussed here; instead, we focus on the economics of the proposed transition to VRE.

The cost of electricity is important for a country's global competitiveness and is a key element for economic development as well as the discussion on energy policy as a whole. Electricity systems are complex, which is also driven by the fact that *a functioning electricity system can supply usable power if, and only if, electricity demand equals electricity supply at all times*, every second. This unique characteristic of electricity systems drives costs. We need to differentiate between cost, value, and price, which are not the same. Below, we discuss only cost.

- **Cost** – the resources and work required for production.
- **Value** – the intrinsic value or utility to the consumer of a particular application as compared to its alternatives.
- **Price** – what consumers or the market are willing to pay. The price is driven by costs, demand, and supply. It is influenced or distorted by government or company intervention, such as laws, mandates, subsidies, etc.

The concept of the true full cost of electricity, FCOE, is introduced in the following chapter. Cost of electricity has been studied in detail by several government organizations and universities. The full cost of electricity (at UT designated as FCe-) was described in a number of white papers published at the University of Texas (UT) 2018. However, UT focuses on transmission and distribution, paying less attention to backup, storage, and the intermittency of VRE[8]. Also, the lower asset utilization of backup systems is not discussed in detail.

The OECD (OECD NEA 2018) references the full cost of electricity, distinguishing between (a) plant-level costs, (b) grid-level system costs, and (c) external or social costs outside the electricity system. The argument is that the full cost must include all three categories, which we agree with. The OECD study pays more attention to the higher volatility and complexity with added VRE in the system, but for instance, energy returns, (eROI) or cost for recycling are not considered. In the OECD's discussion on pollution and GHGs, the life-cycle emission and non-emission impacts of energy systems are also not reflected; the focus is on combustion/operation and mostly on CO_2 (OECD NEA 2018, p101). The study also only makes marginal reference to resource and space considerations. On costs, the following OECD statements are important, and we wholeheartedly agree with them:

- *"When VREs increase the cost of the total system, ... they impose such technical externalities or social costs through increased balancing costs, more costly transport and distribution networks and the need for more costly residual systems to provide security of supply around the clock"* (OECD NEA 2018, p39).

- *"From the point of view of economic theory, VREs should be taxed for these surplus costs [integration costs above] in order to achieve their economically optimal deployment."* (OECD NEA 2018, p39).

Various other electricity cost metrics exist, such as LCOE, VALCOE, LACE, LCOS, Integrations Costs of VRE, etc. In order to provide a complete cost picture, we will conceptually introduce and detail the full cost of electricity (FCOE) to society. FCOE encompasses 10 different categories. This illustrates its complexity, and many of the categories are not easily measurable (see Figure 28). We have not yet found these 10 categories considered in full by any energy-focused economic institution, government, university, private company,

---

8   LCOE = Levelized Cost of Electricity; VALCOE = Value-Adjusted Cost of Electricity; LACE = Levelized Avoided Cost of Electricity; LCOS = Levelized Cost of Storage; VRE = Variable Renewable Energy.

or anywhere in the media. Usually only two or three categories are discussed, and the levelized cost of electricity (LCOE) is the measure erroneously used most often. The socio-economic and environmental benefits of understanding the methods for electricity cost determination are substantial and require further study. FCOE is discussed in the academic research paper *Full cost of electricity 'FCOE' and energy returns 'eROI'* by Schernikau and Smith 2022.

IEA Electricity 2022, p41, clarifies for its value-adjusted levelized cost of electricity, VALCOE, that *"although the VALCOE goes beyond the LCOE and provides a fuller and more accurate measure of competitiveness, it is not all encompassing: it does not yet account for network integration and other indirect costs..."* It must be noted that VALCOE also does not include all environmental costs, only prices for CO_2.

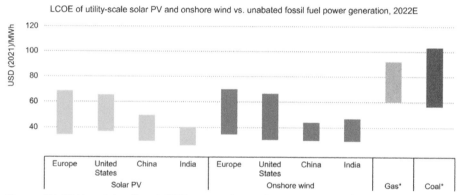

Figure 27: IEA's misleading LCOE comparison of intermittent solar and wind next to dispatchable gas and coal

*Refers to the same regions shown on the graph: Europe, United States, China, and India.

IEA Notes: Gas refers to a combined-cycle gas turbine (CCGT) and coal to supercritical; fuel costs for gas and coal and CO_2 prices reflect the levels projected in the IEA World Energy Outlook 2021 STEPS (Stated Policies Scenario) and do not consider actual spot and forward market prices; variable renewables remain competitive in terms of value-adjusted LCOE (VALCOE) (Box 1.1).

Source: IEA WEO 2022, p37

Unfortunately, as late as June 2022, *the IEA was still misleadingly displaying the LCOE of intermittent variable "renewables" next to dispatchable coal and gas in their most recent World Energy Investment Report* (IEA WEO 2022, p37), while only making reference to VALCOE in a footnote. It was also not made clear until later in the report that VALCOE does not show the full cost either. This continues to delude policymakers who do not have time to go into the details. Looking at a graph such as Figure 27, they draw the erroneous conclusion that *"wind and solar are now cheaper than gas and coal"*. Later, policymakers and politicians cannot understand why the cost of electricity, and with it prices, continues to increase as the share of solar and wind in the system increases.

3.1. Full cost of electricity – FCOE

In economics class, we learned that the question *"What does it cost?"* can have many answers.

The answers depend on why you are asking the question. For instance, the question *"how much does one ton of iron ore cost?"* will have many answers. Are you asking how much one additional ton of iron ore would (marginally) cost when it comes from a mine producing 10 million tons p.a.? Are you asking what the average cash (total) cost per ton of iron ore was last year for that same mine? Or are you asking what the true full cost to society is for an average ton of iron ore, when including all environmental costs as well as the opportunity costs of not having the iron ore? Bear in mind that one mostly uses prices (not costs) for many aspects of calculating the average full cost of a mining project, which – in itself – is incorrect when considering the true cost to society but correct for investment purposes.

Since the question of electricity is one at societal level, or at least at national level, we will attempt to define the true full cost of electricity (FCOE) to society. 10 cost categories determine what we refer to as the **Full Cost of Electricity 'FCOE'** to society:

1. **Cost of Building** electricity generation/processing equipment such as solar panels, a power plant, mine, gas well, or refinery, etc. (often referred to as investment costs, not price).

2. **Cost of Fuel**, such as oil, coal, gas, uranium, biomass, solar, or wind (the latter two are accounted for at a zero cost of fuel). This would include all costs required to make the fuel available, including processing, upgrading, and transporting the fuel through pipelines, on vessels, rail, or trucks. It would also include costs for rehabilitating the source of the fuel, such as mines or wells. LCOE often assumes

that the *price* for CO_2 is part of the *cost* of fuel, but we correctly define a separate category 7. Cost to *Environment*, which includes GHGs. Therefore, the CO_2 price should not be part of cost of fuel.

3. **Cost of Operating** and maintaining the electricity generation/processing equipment.

4. **Cost of (Electricity) Transportation/Balancing** systems to the end user, such as transmission grids, charging stations, load balancing, rectifiers, inverters, smart meters, and other IT technology.

   a) Higher shares of VRE in the system increase the complexity and fragility of transmission, balancing, and conditioning infrastructure. As grids and energy systems become more complex and fragile (see Figure 13), grid transmission infrastructure and control systems become more expensive.

   b) US Midcontinent Independent System Operator MISO 2021 writes: *"... beyond 30% renewable penetration the system as a whole is facing new and shifting risks rather than simply local issues."* IEA 2022, fig. 32, writes: *"Shifting away from centralized thermal power plants as the main providers of electricity makes power systems more complex. Multiple services are needed to maintain secure electricity supply."*

   c) Computerized network-connected control systems are then subject to the threat of and costs of cyber-attacks. Refer to BCG Guide to Cyber Security (BCG 2021a) and the March 2022 cyberattack on satellite infrastructure targeting German windmills (Willuhn 2022). Also refer to the 2017 attack on Ukrainian energy infrastructure described in the excellent book *Sandworm – a new era of cyberwar* (Greenberg 2019).

5. **Cost of Storage**. Storage is always required by "renewable" energy systems, including medium and long-term storage as well as load balancing. That cost must include the cost of building and operating pumped hydro, batteries, hydrogen, etc. Keep in mind that oil, coal, gas, uranium, and biomass have energy storage built in.

   a) The full cost of storage must include, just for storage alone: (1) cost of building, (3) cost of operation, (7) cost to environment, (8) cost of recycling, and (10) other metrics MIPS, lifetime, and eROI.

6. ***Cost of Backup*** technology; electricity systems must include redundancy in case something happens to a power plant or equipment. All reliable electricity systems are overdesigned, usually by ~20% of the highest (peak) power demand. In addition:
   a) Every single VRE installation equipment, such as for wind and solar, requires 100% long-duration backup, storage, or a combination of both as by nature they are not dispatchable or predictable. Modern meteorology models are capable of giving perhaps 75-80% reliability predictions for local wind and solar conditions 24-36 hours in advance.
   b) Conventional power plants are often used as a backup for VRE. The higher the share of VRE in the electricity system, the less the backup capacity will be used, causing lower asset utilization. Thus, the cost of backup increases logarithmically as the VRE share in the energy system increases beyond a certain point (see also IEEJ 2020, p124ff).
   c) Thus, backup capacity may and currently does substitute for long-term storage and is included herein as a separate category since it has a different quality and cost. It is important to avoid double counting.

7. ***Cost to Environment*** includes the true cost (not arbitrary taxes or subsidies) of all emission and non-emission environmental impacts from power generation technology along the entire value chain. This would include but not be limited to life-cycle GHGs from building to recycling the equipment, particulate matters, SOx, NOx, as well as non-GHG climate effects, and non-emission impacts, for example on the local climate, on plant and animal life, or from raw material extraction and recycling (Schernikau and Smith 2022 and Smith and Schernikau 2022).
   a) The climatic and warming effects of large-scale wind and solar installations are well-documented but remain mostly ignored by the industry, policymakers, and investors (see Barron-Gafford et al. 2016, Miller and Keith 2018, Lu et al. 2020, Smith and Schernikau 2022). The concept that solar and wind are free comes into conflict with climate, environmental, and ecological effects. The true "fuel" cost of wind and solar is just now becoming clear as PV parks and wind farms substantially grow in size.

3. COST OF ELECTRICITY AND eROI

b) Other environmental costs of biomass are well-documented and illustrated in the documentary *Planet of the Humans* by Moore 2020, which is sometimes considered controversial.

c) The benefits of CO_2 due to its proven fertilization effects for all plant life would also have to be incorporated as a negative cost to environment (Zhu et al. 2016, NASA 2019, WEF 2019).

d) Environmental costs (other than from GHGs) from fossil fuels need to be fairly evaluated and included.

e) For the cost of global warming, we refer to Nordhaus 2018, Lomborg 2020, and Kahn 2021 as well as this document's Chapter 4.3 *Decarbonization and "Net-Zero"*.

8. **Cost of Recycling**, decommissioning, or rehabilitation of electricity generation and backup equipment at the end of its lifetime. See also *The Hidden Cost of Solar Energy*, published by INSEAD and Harvard (Atasu et al. 2021).

9. **Room Cost** (sometimes called *land footprint* or *energy sprawl*) is a new cost category relevant for low energy density "renewable" energy such as wind, solar, or biomass. Due to the low energy density per m² of wind, solar, or biomass, they take up far more space than conventional energy generation installations, where room costs tend to be negligible, at least relative to VRE. These larger space requirements negatively impact our environment and must be considered since space requirements come into direct conflict with environmental, ecological, and living space needs.

a) Room cost includes direct costs and opportunity costs related to the larger space required and the impact on, for example, sea transportation routes, crop land, forests, urban areas, increasing water scarcity in aridic areas, and noise pollution, etc. (Nguyen et al. 2021 a/b, Roos and Vahl 2021). Double counting needs to be avoided with point 7. Cost to Environment.

b) A new coal power plant in India would require around 2,8 km² per 1 GW installed capacity plus the space for the coal mining (Zalk and Behrens 2018, CEA 2020), which needs to be restored to its previous contours after mining is complete. A new solar park would take around 17 km² per 1 GW installed capacity, plus the space for mining the resources to build solar (wind requires more than double this space). 1 GW installed solar capacity would generate much less electricity due to solar's low capacity factor. Adjusting for a 16,5% average Spanish solar capacity factor, this would translate to a comparable 93 km^2 for solar, or a multiple of 33x compared to coal. Additional space is required for backup and/or storage, as well as overbuilding to "charge" the backup, due to solar's intermittent nature (Schernikau and Smith 2021).

c) The space required, and therefore room cost, per installed megawatt of VRE increases with higher installed capacity. This has to do with the reduced capacity factor for wind in larger wind farms (see wake effect, Smith and Schernikau 2022) as well as the reduced value of additional VRE beyond an optimal penetration level (NEA 2018, p84ff).

10. **Other Metrics**: Three more elements of the full cost of electricity (FCOE) are metrics that are not measured in US dollars but are important for the environmental efficiency of electricity generation. None of these metrics are included in LCOE.

    a) **Material input per unit of service (MIPS)**: measures the material or resource efficiency of building and operating energy equipment in tons of raw materials per MW capacity (Figure 22) and per MWh of produced electricity. MIPS for energy equipment thus measures an important element of environmental impact. The US Department of Energy (DOE) and the IEA have documented the high material input for "renewable" technology and capacity (Figure 22 and DOE 2015, IEA Minerals 2021, p6). Refer to Davie 2022 for the human cost of mining for cobalt in the Congo (DRC). Fossil fuel material input and dangers also need to be fairly evaluated.

    b) **Lifetime**: measures how long equipment is used on average before it is retired or replaced. We need to consider that repowering of wind and solar significantly reduces the designed lifetime. It is not uncommon for conventional power plants to have a lifetime that is three to four times longer than VRE, and wind and solar have shorter lifetimes than designed due to "repowering". Wind turbine rotors are often replaced after 8-10 years, less than half the expected lifetime of a wind turbine.

The removal, recycling, installation, and rebalancing of the rotors on a large wind turbine is a non-trivial task requiring significant downtime.

c) **Energy return on investment (eROI)**: in a way summarizes a large portion of all the measures mentioned above but is not concerned with cost, rather net energy efficiency. EROI also accounts for the energy efficiency of building, operating, and recycling equipment. It includes all embedded or embodied energy. An eROI of 2:1 means investing 1 kWh of input energy for every 2 kWh of output energy. As per Weissbach et al. 2013, solar and biomass in Northern Europe have a buffered eROI of between 2-4 (buffered means including backup or storage). Prieto and Hall 2013 estimate that solar has an eROI of ~2,5 in Spain. Nuclear has an eROI of around 75, and coal and gas around 30, which we consider to be optimistic. Roman culture, the most efficient pre-industrial civilization, reached an eROI of 2:1. Much uncertainty remains about actual eROI values. See next chapter.

FCOE attempts to estimate the true cost to society that is relevant when estimating the global cost of the energy transition in relation to the global cost of any human-caused climatic changes. We emphasize here that the full cost of electricity (FCOE) to society does not include taxes or subsidies. Governments sometimes impose government set prices or taxes in an attempt to emulate such true costs and support research & development, or simply for tax income. FCOE will account for all "true costs" and therefore may not be the right metric for investment decisions that have to incorporate taxes, subsidies, or the prices (rather than costs) of certain elements.

Direct and indirect subsidies for wind and solar or fossil fuels are thus not included (see Footnote 2 on page 29 for explanation). For VRE, indirect subsidies would include the lack of "carbon" taxes, even though the production and recycling of solar and wind capacity and backup systems cause significant GHG emissions. Indirect subsidies for wind and solar would also include the lack of integration costs for VRE, as explained earlier and detailed in OECD NEA 2018, p39.

Please note that to date, CO_2 or "carbon" taxes only include direct CO_2 emissions from fuel combustion, leaving out life-cycle emissions, methane, and other GHGs. Additionally, less than half of anthropogenic CO_2 emissions end up airborne, with the remainder being taken up by nature (see

IPCC, as detailed in Schernikau and Smith 2022 on *Climate Impacts of Fossil Fuels*). Incorrectly, this is not considered in any "carbon" taxation scheme. Therefore, **CO_2 taxes are misleading and wrong, leading to undesired economic and environmental distortions such as the switch from coal to gas** for "climate reasons", dismissing the higher climate impact of methane emissions associated with gas and especially LNG production, other climate forcings (Mar et al. 2022, p133, Dreyfuss et al. 2022), as well as any so-called "scope 3" emissions of "renewables" and electric vehicles.

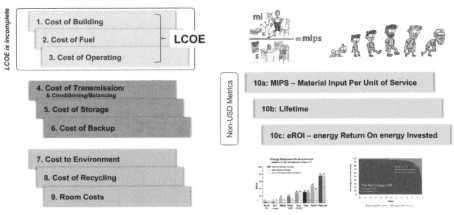

Figure 28: Full cost of electricity (FCOE) to society – a complete picture

Note: Age cartoon original by Alexandra Martin; energy cliff from eROI for beginners; MIPS cartoon from Seppo.net, eROI from Weissbach et al. 2013.

Source: Schernikau Research and Analysis

From the above analysis, it can be concluded that the levelized cost of electricity (LCOE) – which only includes the cost of building (1), cost of fuel (2), cost of operation (3), and sometimes certain CO_2 taxes (part of the cost to environment 7) – is neither a reliable nor an environmentally or economically viable measure with which to evaluate different forms of energy generation at a national or societal level. Only FCOE includes all relevant economic and environmental costs from emissions and non-emissions, though its true value is difficult – but not impossible – to determine. Since no one has yet calculated the true FCOE, we are not able to compare FCOE to LCOE other than stating that FCOE is higher than LCOE. Calculating FCOE is complex and a larger task requiring financial and human resources. The cost increase from LCOE to FCOE is significantly more for variable "renewable" energy than for conventional energy.

Renowned energy think tanks such as the International Energy Agency (IEA) in France, the Institute of Energy Economics (IEEJ) in Japan, the OECD, and the US Energy Information Agency (EIA) *have pointed out the incompleteness of LCOE multiple times. Yet LCOE continues to be widely used despite its failings* (IEA WEO 2022, p37), usually without clear disclaimers or notes, even by these agencies themselves, as well as by governments, banks, institutions, NGOs, companies, large consulting firms, many scientists, and the press.

Undesirable effects occur when conventional fuels and variable "renewable" energy VRE (wind and solar) are mixed to provide a country's electricity. These effects could be measured in their entirety by FCOE categories 1-10 as listed above. For instance, beyond a small capacity share, the cost to a nation's electricity system always increases with higher shares of variable "renewable" energy VRE, such as wind and solar (IEEJ 2020, p124ff, IEA 2019, and IEA Electricity 2020, p13). The cost increases include but are not limited to the previously discussed differential energy density and efficiency, intermittency and thus backup/storage requirements, low capacity factors, interconnection costs, material and energy costs, low eROI, efficiency losses of backup capacity, room costs for the space required and plant/animal life destroyed, recycling needs, and so forth.

- The IEA confirmed in December 2020 (IEA Electricity 2020, p14): *"... the system value of variable renewables such as wind and solar decreases as their share in the power supply increases."* This would also remain true if the price of "renewable" capacity (cost item 1: Cost of Building) continued to decline or were even to reach zero. For example, this conclusion would not change even if the price of solar panels produced with coal power in China partially using forced labor were to reach zero (Murphy and Elima 2021). This would also remain true if wind or solar technology were to reach a quantum efficiency of 100%, which would be impossible.

Learning curves such as illustrated by JP Morgan 2022, p39, correctly estimate marginal cost reductions for energy technology stemming from experience curves and technological advances. Learning curves are a long-standing concept, already discussed in detail by psychologist Hermann Ebbinghaus in 1885, and are used as a way to measure production efficiency and forecast costs (Investopedia 2020). The learning curve is usually depicted in some sort of S-shape, meaning that the impact of learning curves declines the more established a technology or system becomes. When it comes to cost reduction projections for "renewable" technologies, one must look at two things: (1) the material or raw material input requirement making up a large portion of costs, and (2) the irrelevance of any projected reduction of marginal costs when variable "renewables" need to be integrated into the energy system. Therefore, the learning curve benefits are true and correct on a marginal cost basis, but their importance diminishes as the share of variable "renewable" energy in our electricity system becomes larger, as illustrated in this chapter. Also, raw material and energy input play an important role in "renewable" costs, as illustrated during 2022, when "renewable" costs reversed past downward trends. *The logarithmic increase in variable "renewable's" integration costs and low integration costs as well as low eROI, rather than their reduced production costs, are the real cost driver.*

LCOE is inadequate to compare intermittent forms of energy generation with dispatchable ones, and therefore also when making energy policy decisions at a national or societal level. However, LCOE may be used selectively to compare dispatchable generation methods with similar material and energy inputs, such as coal and gas. Using FCOE, or the full cost to society, wind and solar are not cheaper than conventional power generation and in fact become more expensive as their penetration of the energy system increases. This is also illustrated by the high cost of the so-called "green" energy transition, especially to poorer nations (see Chapter 4.2 and Figure 38, McKinsey 2022a and Wood Mackenzie 2022). If wind and solar were truly cheaper – in a free market economy – they would not require trillions of dollars of government funding or subsidies, or laws to force their installation and utilization.

3.2. Energy return on energy invested – eROI

The environmental efficiency of our energy systems is more complex than CO_2 emissions alone. In particular, net energy efficiency or energy return on energy invested *(eROI), material input, lifetime, and recycling efficiency need to be considered as they determine very important additional environmental and economic elements for evaluating electricity generation.* Modern society requires a minimum inherent net energy efficiency (eROI) to function. Different activities have different minimum energy efficiency requirements, as illustrated in Figure 29. More advanced societal activities require a higher net energy efficiency (eROI). Flying to the moon may be one of the most advanced activities requiring highest possible net energy efficiency and energy density to become possible.

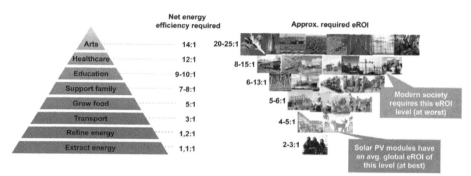

Figure 29: Advanced societies require high net energy efficiencies (eROI)
Source: Adapted from Prieto and Hall 2013

eROI measures the net energy efficiency of an energy-gathering system. It does not count "natural" energy input such as from the sun or wind, or "natural" embodied energy inherent in extracted coal, gas, oil, or uranium. It counts the input energy required to make usable electricity or a fuel product available for consumption, and then calculates the ratio of usable output energy and the required input energy. EROI accounts only for energy production, not for energy consumption.

Higher eROI translates to lower environmental and economic costs, thus lower prices and higher utility. Lower eROI translates to higher environmental and economic costs, thus higher prices and lower utility. When we use less input energy to produce the same output energy, our systems become environmentally and economically more viable. When we use a relatively high amount of input energy for each unit of output energy, we risk what is referred to as "energy starvation". At an eROI of 1 or below, our systems are operating at an energy deficit; at ~2:1, we have matched Rome's energy efficiency from 2.000 years ago.

> *Note: Vaclav Smil's "Energy and Civilization – a History" (Smil 2017) is an excellent, highly acclaimed book on the subject of energy. In addition, we recommend Kiefer 2013 and Delannoy et al. 2021 for more detailed discussions on eROI. Kis et al. 2018 approach eROI by using GER*

(Gross Energy Ratio) and GEER (Gross External Energy Ratio). Kis et al. define GEER as life-cycle eROI and find a global GEER average of approx. 11:1. Even Wikipedia has a separate page on Energy Return on Investment. Due to the complexity of eROI, more research is required to harmonize the approach for its determination. See Schernikau's proposal to calculate eROI adapted from Prof. Michaux 2021 (Figure 31).

Carbajales-Dale et al. 2014 calculate that the average solar PV can only "afford" 1,3 days of battery storage from a net energy efficiency point of view "before the industry operates at an energy deficit". As per the researchers, wind, from a net energy efficiency point of view, can "afford" over 80 days of geological storage (12 days of battery storage). We disagree with the calculations of Carbajales-Dale et al. 2014 and find them too optimistic because they make certain unrealistic assumptions. For the mentioned net energy efficiency calculations, the researchers made the simplifying yet unrealistically positive assumption that a generation technology is supplied with enough energy flow (either wind or sunlight) to deliver 24 hours of average electrical power output every single day. This means that days or weeks with no sun or wind would multiply the storage requirement and therefore further diminish the net energy efficiency or eROI (see Chapter 2.4 *Energy storage*). Carbajales-Dale et al. included the proportion of electricity output consumed in manufacturing and deploying new capacity.

Even without having done the detailed bottom-up calculation due to lack of funds and time, it can be concluded based on available research that **wind and solar have a low eROI and are therefore a step backward in history in terms of system energy efficiency. Their grid-scale employment risks energy starvation and is therefore neither economically nor environmentally desirable.** We would like to point out that for certain applications, i.e., for electricity or heating water in remote unconnected villages or heating a pool not connected to the grid, solar and wind may be a desirable complement to our energy systems. The installation of wind and solar does reduce the amount of fossil fuels combusted, assuming no increase in power demand, which is the only positive of their employment. This positive aspect comes with large costs, as summarized illustratively in Figure 8: Summary of shortcomings of variable "renewable" energy for electricity generation.

In addition, biomass has significant negative environmental effects and a low eROI, as detailed in the documentary *Planet of the Humans* by Moore

2020 and Weissbach et al. 2013. The US Energy Protection Agency writes on the *Economics of Biofuels* (EPA 2022):

> "... many biofuel feedstocks require land, water, and other resources, research suggests that biofuel production may give rise to several undesirable effects. Potential drawbacks include changes to land use patterns that may increase GHG emissions, pressure on water resources, air and water pollution, and increased food costs. Depending on the feedstock and production process and time horizon of the analysis, **biofuels can emit even more GHGs than some fossil fuels on an energy-equivalent basis**. Biofuels also tend to require subsidies and other market interventions to compete economically with fossil fuels, which creates deadweight losses in the economy."

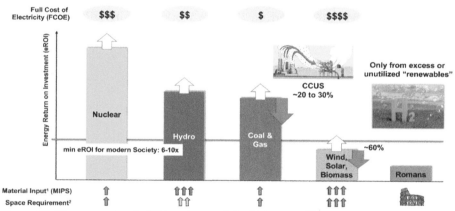

Figure 30: *The concepts of eROI and material efficiency – illustrative*

Note: White arrows illustrate future technological improvements; red arrows illustrate loss of energy and therefore loss of eROI from CCUS or "green" H_2 systems.

(1) Material Input MIPS measures the resource efficiency, i.e., material input required per unit of output, here for example per MWh of produced electricity, which includes the materials required to build the generation capacity.

(2) Space requirement measures the land footprint per unit of electricity produced.

Source: Schernikau Research and Analysis

The industrial revolution reduced humanity's dependency on biomass, hydro, and wind. Based on the newfound **high-eROI-coal-energy**, this energy revolution allowed for a dramatic increase in standards of living, life expectancy, and industrialization, as well as a decrease in heavy human labor and the abandonment of slavery. This revolution and its positive impact on human life was only possible due to a drastic increase in energy availability and energy efficiency, or eROI. The energy revolution involved diversifying away from 100% biomass burning toward fossil fuels, hydro, and later nuclear.

Prior to the industrial revolution, human development peaked during the Roman Empire with an estimated sustained eROI of around 2:1[9]. During the 20th century, petroleum's high eROI, higher energy density, and versatility enabled the transportation revolution with its cars, aircraft, and rockets. To appreciate the magnitude of petroleum's discovery, consider that *three tablespoons of crude oil contain the equivalent of eight hours of human labor* (Kiefer 2013 and Footnote 9). Figure 30 schematically illustrates the concept of eROI in today's electricity systems and the impact of CCUS or hydrogen storage on net energy efficiencies (Supekar and Skerlos 2015).

Dr. Euan Mearns 2016, based on Kiefer's work, explains eROI and points out that modern life requires a minimum eROI of 5-7, while *most solar and many wind installations have an eROI of below 5, depending on location, and are therefore inherently energy insufficient when it comes to supporting society at large. As per Prieto and Hall 2013 (p115), solar in Spain has an estimated eROI of ~2,5, just above the average Roman eROI.* As per Weissbach et al. 2013, solar and biomass in Northern Europe have a buffered eROI of around 2-4. Nuclear has an eROI of around 75, and coal and gas around 30, which we consider too optimistic. Kiefer defines "*The Net Energy Cliff*", which demonstrates how – with declining eROI – society would commit ever-larger amounts of work to energy-gathering activities.

One example is employment; below an eROI of 5-7, such large numbers of people would be working for energy-gathering industries that there would not be enough people left to fill all the other positions our current altruistic society requires. However, some may argue that this is desirable due to the expected long-term threat to human labor posed by artificial intelligence.

---

9   Based on Kiefer 2013: eROI for humans and oxen as the ratio of max. work output divided by food calorie input, calculated from Homer-Dixon's online data as 0,175:1. EROI for Roman wheat, as the ratio of food calorie output divided by labor and seed grain inputs, was 10,5:1. EROI for alfalfa was 27:1. Humans eating wheat yield a heavy labor eROI of 0,175 x 10,5 = 1,8:1. Oxen eating alfalfa yield an eROI of 0,175 x 27 = 4,7:1. Teaming humans with oxen and applying reductions for idle time and for light work/skilled labor versus heavy labor gives ~4,2:1 peak eROI and ~1,8:1 sustained eROI.

3. Cost of electricity and eROI

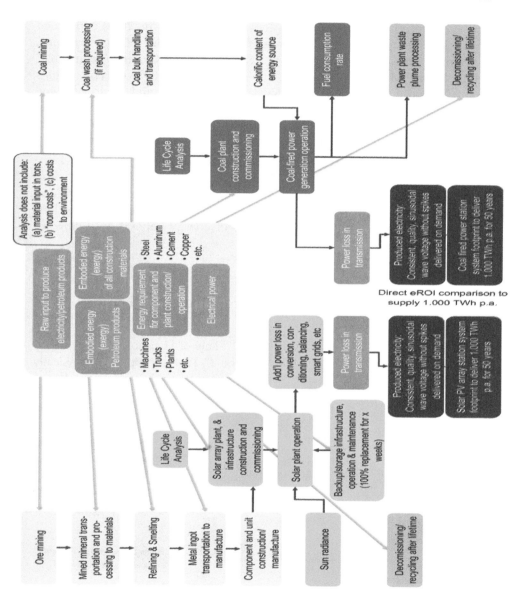

Figure 31: Example of proposed eROI study to compare coal and solar PV

Note: Prof. Michaux proposed comparing the true eROI for a 1.000 TWh annual net electricity supply to the consumer. The analysis was enhanced by Schernikau to include recycling and further detail on backup or storage requirements for solar.

Source: Schernikau Research and Analysis based on Michaux 2021

The IEA's recent World Energy Outlook (IEA Investments 2021) confirmed that global employment would rise from "renewable" energy systems, therefore providing evidence for the lower eROI of "renewable" technologies. McKinsey 2018, not considering the eROI concept, argues that automation will replace low-level workers; this trend is already well underway. Those without higher technical and intellectual skills may become unemployable in the future workforce. In effect, McKinsey sees enough labor available for future more labor-intensive energy systems.

The principle of energy return on investment (eROI) is at the core of society's energy efficiency, which is at the core of humanity's development and survival. Today's knowledge of eROI appears insufficient, and more exact numbers cannot be given yet. We propose a global scientific effort to estimate eROI for all forms of energy generation, following the logic illustrated in Figure 31.

3.3. The 2nd Law of Thermodynamics' impact on energy systems

The preface already introduced the 1st and 2nd Law of Thermodynamics. Figure 32 tries to summarize the laws' function. The 1st Law is simple as it basically states that energy can never be lost, only converted from one form to another.

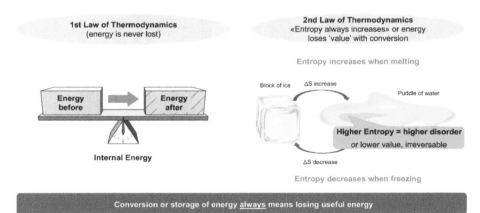

Figure 32: 1st and 2nd Law of Thermodynamics in closed systems
Source: Schernikau Illustration, graphs from (https://bit.ly/3B2SU6D)

The 2nd Law introduces the concept of entropy, which can be used to indicate the usefulness or value of energy (high entropy = high disorder or low value of energy). Essentially, the 2nd Law explains why, in a natural state, heat always moves from warm to cold and not the other way around. When energy is converted from one form to another, entropy always increases and "useful" energy is lost. The logical conclusion for our modern energy systems is that *we need to avoid the conversion and storage of energy, as well as increasing the complexity of our energy systems as much as possible*, as all of these result in the loss of useful energy (Figure 33). For exactly these reasons, "renewable" energy systems result in a loss of usable energy, which civilization cannot allow if the conversion reduces eROI below the sustainability threshold for civilization.

This loss of useful energy is critically important because it directly translates into reduced system energy efficiency. The direct result of converting wind power to hydrogen, storing and transferring hydrogen, and converting hydrogen back to power is a lower eROI. It also directly results in the warming of our biosphere because the lost energy is never actually lost and essentially appears in higher entropy or, equivalently, low-value heat. The net efficiency of a gas- or coal-fired power plant is a direct result of the 2nd Law of Thermodynamics. Every process that takes place in the boiler, the turbine, or the generator "costs" energy that is lost in the form of low-value/waste heat to our surroundings.

We have already established in this chapter that the "green" energy transition to variable "renewable" energy in the form of wind and solar does already and will continue to substantially increase the cost of electricity. We will establish in Chapter 4.2 that this rise in cost will primarily burden poorer people and developing nations (McKinsey 2022a, Wood Mackenzie 2022, and Figure 38). With the concept of the 2nd Law of Thermodynamics, we have identified one more major reason why *the "green" energy transition can only reduce global net energy efficiencies because it requires more complex energy systems and increases storage, conversion, and transmission losses*. The IEA summarized the issue of increasing complexity in their article *Energy transitions require innovation in power system planning* (IEA 2022, see Figure 33) as follows:

- *"Shifting away from centralized thermal power plants as the main providers of electricity **makes power systems more complex**. Multiple services are needed to maintain secure electricity supply.*
- *In addition to supplying enough energy, these include meeting peak capacity requirements, keeping the power system stable during short-term disturbances, and having enough flexibility to ramp up and down in response to changes in supply or demand."*

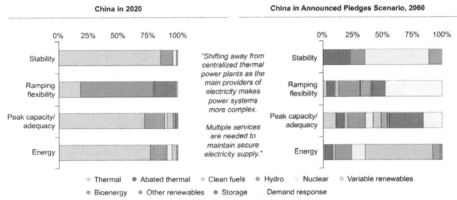

Figure 33: Current and future energy security in China
Source: Based on IEA 2022

The 1ˢᵗ Law of Thermodynamics states that our produced and consumed energy ends up in low-value or high-entropy heat and thus warms our biosphere, adding to the measured temperature increase (for the amount, see also Soon et al. 2015). The total annual energy consumption by humanity is more than 4.000 times smaller than the annual input of energy by the sun, and thus seemingly negligible in total; still, it contributes to the measured temperature increase in and around populated, energy-consuming areas. There is also embodied energy in all products that we produce, which is not immediately released in the form of heat. These products also represent an increase in entropy since recycling to the original elements has high energy cost. In other words, all human existence and production can be viewed as a step toward higher entropy. The heat island effect, as mentioned in Chapter 4.3, is also a manifestation of the heat emitted to our surroundings and is a regional consequence of urbanization.

When we produce energy from sources such as nuclear, oil, coal, gas, or even geothermal, then we are taking energy that is "inside our planet" and in the end converting it into low-value heat that regionally contributes to warming our biosphere. When we use this energy from solar radiation by employing photovoltaics or wind turbines, the only way we could avoid a "net" warming of our planet would be to disregard the warming from solar panels' absorption, shifting atmospheric circulation, and other regional climate impacts (Barron-Gafford et al. 2016, Lu et al. 2020, Miller and Keith 2018, Smith and Schernikau 2022). Additionally, solar PV and wind energy could only avoid a "net" warming of our planet if we were to disregard the energy used for building and recycling the equipment or systems required to extract and use solar energy. High CO_2-emitting forms of "producing" energy, such as coal or gas, partially offset the warming of the biosphere through fertilization and greening, which reduces solar warming. Solar radiation can only do one thing: grow a plant or warm the Earth (see Schernikau and Smith 2022).

4. The projected future of energy and "decarbonization"

To allow for a "clean energy transition", The Boston Consulting Group (BCG 2021b) projects global wind and solar power capacity to increase at a similar level to the overbuilding that has taken place in Germany over the last 20 years (see Figure 7 and Figure 34). In 2020, global power generation capacity totaled around 8.000 GW, of which over 1.400 GW were wind and solar. In March 2022, 1.000 GW of global installed solar capacity was celebrated in the press (pv-Magazine 2022). In eight years (at the time of writing), that is, by 2030, BCG projects that wind and solar alone will have to reach 8.600 GW, doubling today's entire global electricity capacity. BCG projects that this doubling of **global variable "renewable" energy capacity** has to happen in those few years, in the same way as happened in Germany during the 20 years from 2002 to 2021 (Figure 7).

Based on 2021 IRENA outlook data, BCG also forecasts that global installed wind and solar capacity must reach 22.000 GW by 2050, almost quadruple today's entire global electricity generation capacity. We conclude from our analysis that *these nameplate forecasted capacities will not be reached as the world would run out of energy, raw materials, space, and money* before this could happen, and if they were reached, the economic and environmental impact to society would be close to devastating, as explained in this book.

Such dramatic expansion of wind and solar capacity will result in more fragile and expensive energy systems. It will also negatively impact the environment (see space requirements, backup, material input, eROI, recycling needs, local climate impacts, etc.), offsetting any desired – modeled – positive effects on the global climate from projected anthropogenic GHG-emission reductions. Additionally, total input energy required would rise significantly for the same usable energy available to humanity for final consumption.

On the positive side, and in our view the only positive aspect, is that such wind and solar expansion could reduce the use of fossil raw materials mined if energy demand did not increase; however, this is not realistic. The real question is whether it would truly reduce total raw material use when honestly and accurately accounting for the entire life cycle, beginning with resource mining and encompassing material transportation, processing, manufacturing, and operation, as well as recycling (Figure 22 and Figure 39). Based on our research, we conclude it would not. The environmental impact of the "new" raw materials required for the "energy transition" is exemplified by the rush for cobalt in the Congo (DRC), which has revealed the human cost of the world's "green" energy future (Davie 2022), as well as the dominance of China when it comes to the processing of raw materials and manufacturing of "green" energy products such as solar PV, wind systems, EVs, and more.

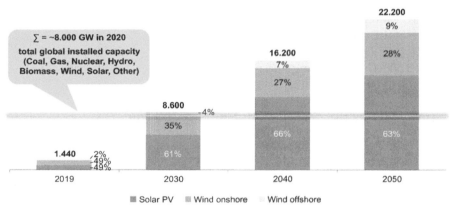

Figure 34: Wind/solar capacity forecast for 2050 to be almost 4x today's total capacity Source: Schernikau Research and Analysis based on IRENA 2021 and BCG 2021b

4.1. Primary energy (PE) growth until 2050

After having risen from ~2 billion to ~8 billion over the past 100 years, the UN projects that global population will rise further from the current ~8 billion to ~10 billion by 2050 (OurWorldInData 2021). The population may peak at around 11-12 billion by the end of the century. Despite continued improvements in energy efficiencies, rising living standards in developing nations are forecast to increase global average annual per capita energy consumption from ~21.000 kWh to ~25.000 kWh by 2050 (Lomborg 2020, BP 2019).

4. THE PROJECTED FUTURE OF ENERGY AND "DECARBONIZATION"

As a result, and as illustrated in Figure 35, *global primary energy consumption could rise by 50% by 2050* (~25% population increase and ~20% PE/capita increase). Energy demand growth is fueled by developing nations in Asia, Africa, and South America. Developed nations are expected to consume less energy in the decades to come, driven by population decrease/stagnation and efficiency increases. However, historically, energy efficiency improvements have always increased energy demand (see *Jevons Paradox*, Polimeni et al. 2015). For illustration purposes, we highly recommended the book *Life After Google* (Gilder 2018), which explains the increased energy requirement for global computing. We also draw your attention to one of the largest planned hydrogen production facilities in Texas; it is designed to meet future fuel demand for Elon Musk's SpaceX program rather than to replace existing fossil fuel demand (see Chapter 2.5 and Collins 2022a).

We reiterate that recent models by McKinsey estimate that global primary energy demand will only increase by 14% by 2050, while the IEA's 2021 and IRENA's 2022 "Net-Zero" pathways model a primary energy reduction of up to ~10% by 2030, that is, in 8 years from the writing of this book. However, we, together with the energy industry at large, question this (IEA Net-Zero 2021, McKinsey 2021a). Adhering to these pathways and the reductions they require would necessitate severe economic restrictions, increased global poverty, global population declines, and unnecessary human suffering. First signs of energy starvation could be witnessed in Europe during 2022 with large industrial operations forced to close.

The same reports estimate that global electricity generation will almost double from 2020 to 2050, also being driven by the projected electrification of transportation. The Institute of Energy Economics, Japan (IEEJ 2021) predicts global primary energy demand to increase by 30% by 2050, while the American EIA predicts a ~50% increase that is in line with our own projections (EIA 2021). Kober et al. 2020 compare various energy scenarios and point out that essentially all energy scenarios assume a decoupling of economic growth and energy consumption in the future. *"All the examined scenarios show a strong rise in power generation that exceeds the rate of growth in primary energy consumption."* We assume that Kober et al. did not consider eROI, material input, or the Jevons Paradox.

Growth in electricity demand will surpass primary energy growth, partially due to the global electrification of operations. Electricity's share of primary energy will also increase because our lives are becoming more computerized and "gadgetized". Electricity is planned to replace significant non-electricity energy consumption for transportation (i.e., EVs), heating (i.e., heat pumps), and industry (i.e., DRI for steel production), which we have shown will reduce net energy efficiencies if wind, solar, or biomass are used.

- Dierig 2022, from the leading German newspaper *Die Welt*, quoted Thyssen Steel Chairman Osburg in February 2022: *"going climate neutral will increase energy demand 10x from 4,5 TWh to 45 TWh"* for Europe's largest steel plant, in Duisburg, alone. This additional power demand equals approximately four times the annual electricity demand for the city of Hamburg or around 8% of Germany's total electricity consumption, just to produce the same amount of steel Thyssen had previously produced using fossil fuels.

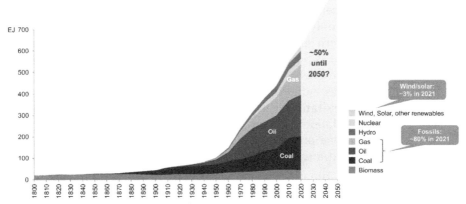

Figure 35: Global primary energy from 1750 to 2050
Note: Primary electricity converted by direct equivalent method.

Source: Schernikau Research and Analysis based on data compiled by J. David Hughes. Post-1965 data from BP, Statistical Review of World Energy (https://on.bp.com/3Rabcs9). Pre-1965 data from Arnulf Grubler 1998: Technology and Global Change: Data Appendix (https://bit.ly/3Rph4xE), and World Energy Council 2013: World Energy Scenarios. Composing energy futures to 2050 (https://bit.ly/3RtpzYz)

Despite hoped-for technological improvements, *it is prudent to assume that wind and solar will not be able to generate enough total energy to match the expected demand increase by 2050*. This is confirmed by the IEEJ 2021, which forecasts an absolute increase in fossil fuels' share in primary energy by 2050 in its reference case. In July 2021, the IEA confirmed that *"...[renewables] are expected to be able to serve only around half of the projected growth in global [electricity] demand in 2021 and 2022"* (IEA Electricity 2021). The "renewable" share will only be a fraction of primary energy growth, perhaps 20%, since today around 40% of primary energy is used for electricity production (Figure 3). The strong post-Covid economic recovery year 2021 evidenced that it required fossil fuels, namely coal during that year, to fuel the economic growth (IEA Electricity 2022).

Even if wind and solar could fulfill all future increases in primary energy demand, it is evident that for the next 30 years and beyond we will continue to depend on conventional energy resources for most, if not the vast majority, of our global energy needs.

For recent "Net-Zero" pathways (IEA Net-Zero 2021, IRENA 2022) and scenarios to succeed on paper, they require a number of highly optimistic, even unrealistic, assumptions related to rapid advances in technology development, hydrogen penetration, demand curtailments, raw materials with controllable prices and supply availability, and so forth. They also largely dismiss eROI, material input, lifetime, transmission infrastructure, and realistic recycling assumptions, and thus largely ignore "renewables'" negative economic and environmental impacts. Such models do not account for any additional cost from energy shortages directly resulting from the move to intermittent low eROI wind and solar. The problem that increasing the proportion of "renewables" will cost the poor more and reduce energy security is discussed next.

4.2. Energy shortages and their impact on prices and economic activity

The logical economic conclusions from the previous chapters are that:
1. Future energy requirements outstrip "Net-Zero" pathways and any probable "renewable" generation.
2. "Renewable" electricity will remain a scarce resource, and its widespread adoption would reduce humanity's net energy efficiency below the level critical to sustaining our present advanced civilization.
3. No viable, long-term, grid-scale energy storage solution has yet been found.

Thus, as long as "renewables" remain scarce, any NGO or conglomerate claiming that all its energy growth originates from "renewable" energy such as hydro must realize that one consequence of this is that another consumer who may otherwise have used this hydro energy can now no longer do so. Using a scarce resource for growing one's own business takes away that resource from others. Evidently, the net effect to society is independent of who uses the "green" energy available. In addition, the *loss of "renewable" energy due to conversion for storage or transportation is unavoidable and must be minimized, as is consistent with physics and economics.*

The apparent energy shortage in Europe and other parts of the world, which started in 2021, illustrates the low net energy efficiency (eROI) and the explained high cost of variable "renewable" energy. The lack of investment in conventional forms of energy has resulted in undersupply, while at the same time wind and solar have not been able to satisfy increased demand. Germany's consumer power prices, the highest in any industrialized nation, represent further evidence of the full cost of electricity (FCOE) and are also driven by the relatively high penetration of "renewables". In June 2022, Germany pushed the G7 for renewed investment in LNG, against IEA, UN, and World Bank suggestions to stop all fossil fuel investment. This is good and necessary. It is also only logical, as the repercussions from running out of energy are far worse for our lives, especially for the poor, than any modeled impact from climate change by 2100 (Bloomberg 2022b).

BCG and the International Energy Forum (IEF) had already warned in their December 2020 energy report *Oil and Gas Investment in the New Risk Environment* that *"... by 2030, investment levels [in oil and gas] will need to rise by at least US$ 225 billion from 2020 levels to stave off a crisis"* (BCG and IEF 2020). The press started to pick up this subject in the third quarter of 2021 when energy resources and electricity prices soared and the first signs of global energy shortages surfaced. Investment in coal is pro rata even lower than in oil and gas (Figure 4).

When looking at steel infrastructure, Wido Witecka, a steel expert from Germany's "climate protection" think tank Agora Energiewende, pointed out correctly that the 2020s are a critical decade, since more than 70% of all steel ovens will have reached end of life by the end of the 2020s and reinvestment decisions are required soon (DW 2022). The US, according to independent analyst Paul Sankey, is "structurally short" on refining capacity for the first time in decades; energy experts predict a refinery supply crunch as 700 refineries worldwide are not sufficient to meet demand and investment is insufficient (Sinicola 2022a/b). Energy expert Stein 2022 goes further and points out that *"Energy shortages and inflation will be the new norm as refinery closures outpace construction"* globally. Kearney 2021 points out that 20% of refineries, that is, almost 140, are expected to close worldwide in the next five years. Amid increased ESG regulation, these refineries are unlikely to be replaced easily, especially in the West.

The 2022 Russian invasion of Ukraine also illustrates the fragility of global energy systems and how intertwined energy and politics are, especially when it comes to oil, gas, and nuclear. Access to affordable and reliable energy should not be about politics. Unfortunately, energy policy has been repeatedly misused by both sides of the political agenda. Of the dispatchable forms of energy, coal, hydro, and geothermal energy are the least political. **Wind, solar, and EVs are also political because raw materials and raw material processing are very concentrated in a few countries, particularly China** (IEA Solar 2022 and Figure 21). Below is a list[10] of selected press articles on the topic of the "new energy crisis", which started in 2021 before Putin's invasion; for links, see Footnote 10.

1. *"The world has never witnessed such a major energy crisis in terms of its depth and its complexity"*, IEA Executive Director Fatih Birol said in July 2022 at a global energy forum in Sydney. *"We might not have seen the worst of it yet – this is affecting the entire world."*

2. Bjarne Schieldrop, chief commodities analyst at SEB, March 2022: *"The global economy is facing energy starvation right now and demand destruction will set a limit to the upside eventually."*

3. Vaclav Smil wrote in February 2022, referring to the Russian invasion of Ukraine: *"This war will have many long-term consequences, but possibly none more important than its effects on the future of the European energy supply."*

4. The N24 wrote in February 2022: *"The worst energy crisis since 1973."*

---

10 Sources in order: (1) Bloomberg 11 Jul 2022 (https://bit.ly/3Rpmn02), (2) Telegraph 2 Mar 2022 (https://bit.ly/3eeDEuv), (3) Vaclav Smil, 28 Feb 2021 (https://bit.ly/3Q1k2Hj), (4) The N24, 24 Feb 2022 (https://bit.ly/3q22hgu); (5) CNN, 18 Nov 2021 (https://cnn.it/3KylmiR), (6) Wikipedia 5 Dec 2021 (https://bit.ly/3CNBoVs), (7) Bloomberg, 5 Oct 2021 (https://bloom.bg/3CJnbZl); (8) Globe and Mail, 1 Oct 2021 (https://tgam.ca/3edMfxm); (9) Bloomberg, 18 Sep 2021 (https://bloom.bg/3wL49y5); (10) Nikkei Asia, 27 Sep 21 (https://s.nikkei.com/3Q4cjrZ)

5. CNN wrote in November 2021: *"... anti-poverty organizations and environmental campaigners have warned that millions of people across Europe may not be able to afford to heat their homes this winter ..."*
6. Wikipedia set up a separate page referencing the 2021 Global Energy Crisis in November 2021: *"The 2021 global energy crisis is an ongoing shortage of energy across the world, affecting countries such as the United Kingdom and China, among others."*
7. Bloomberg wrote in October 2021: *"The world is living through the first major energy crisis of the clean-power transition. It won't be the last."*
8. The Globe and Mail wrote in October 2021: *"India's coal crisis brews as power demand surges, record global prices bite."*
9. Bloomberg wrote in September 2021: *"Europe is short of gas and coal and if the wind doesn't blow, the worst-case scenario could play out: widespread blackouts that force businesses and factories to shut. The unprecedented energy crunch has been brewing for years, with Europe growing increasingly dependent on intermittent sources of energy such as wind and solar while investments in fossil fuels declined."*
10. Nikkei Asia wrote in September 2021: *"Key Apple, Tesla suppliers halt production amid China power crunch."* Bloomberg follows in the same month that *"China may be diving head-first into a power supply shock that could hit Asia's largest economy hard just as the Evergrande crisis sends shockwaves through its financial system."*

The human and economic costs from shortages in electricity supply are apparent from countless examples worldwide. A European example includes the Italian power outages on 28th September 2003. On that day, the north of Italy experienced an outage of up to three hours and the south (Sicily) of up to 16 hours. A loss of 200 GWh to customers resulted in an estimated EUR 1,2 billion economic loss (Baruya 2019, former IEA Clean Coal Center). Baruya summarizes: *"In developing regions, such as sub-Saharan Africa, shortages in energy supplies impede business and economic growth. In advanced economies, failure in the power grid and generating capacity has also led to measurable economic losses, such as those seen in Italy in recent years."* Another direct impact of electricity outages will be the loss of human lives and health. It must be noted that none of the "Net-Zero" models or scenarios account for any cost resulting from energy shortages, or worse, energy starvation.

We have shown why *the "energy transition" to variable "renewable" forms of energy such as wind and solar will result in higher electricity costs.* Energy-transition-supporting strategy consultant McKinsey 2022a confirms and also summarizes: *"A Net-Zero transition would have a significant and often front-loaded effect on demand, capital allocation, costs, and jobs."*

Research shows that a rise in electricity prices and energy shortages impacts economic output and hurts the poor. This is also acknowledged by "green" energy transition supporters. Germany's Economy and Climate Minister Robert Habeck (Green Party) said in June 2022, after commenting on the gas and energy shortage: *"Companies would have to stop production, lay off their workers, supply chains would collapse, people would go into debt to pay their heating bills, those people would become poorer."* (Reuters 2022a).

Based on scientific research, Baruya 2019 summarized the impact of rising electricity costs to industries in China, the US, Russia, Mexico, Turkey, and Europe. The coefficients of elasticity between economic output and electricity prices were irrefutably negative. Output declined faster in the non-metallic minerals (cement) sector, metal smelting and processing, the chemical industry, and mining and metal products. For example, in Vietnam, the impacts of an increase in the electricity tariff on the long-run marginal cost of products manufactured using electricity-intensive processes were examined in 2008. An increase in tariffs drove price inflation for all affected goods and services (Baruya 2019).

Baruya 2019 continues and confirms our analysis of how the retirement of fossil-fuel-fired power plants without adequate, reliable, and affordable alternatives will *"reduce the amount of backup power to less than the amount required to meet capacity shortages during peak electricity demand"*. Developing and industrializing nations, such as India, Indonesia, Vietnam, Bangladesh, and Pakistan, will be negatively affected by the cessation of funding from Western financial institutions. Alternative funding may lead to the adoption of less efficient generating technologies resulting in an increased environmental burden. Consequently, industrializing countries that do not invest in high-efficiency, low-emission (HELE) conventional fuel technologies could face higher costs of generation, reducing their competitiveness and as a result slowing economic growth.

The situation cannot make any clearer the need for governments to act and adjust their taxation, subsidies, and energy policies. *If investment in fossil fuels and nuclear does not increase substantially and very soon, a prolonged global energy crisis will be difficult to avoid during this decade.* This would remain true even if all sustainability goals were to be achieved and wind and solar capacity continued to increase as planned or hoped for. Global energy markets during the 2021 Covid-19 recovery period in Europe and Asia and the Russian/Ukrainian war in 2022 are testimonies to the impact of energy shortages. Keep in mind that there is no shortage of energy resources, only a shortage of energy raw material production and reliable electricity generation, driven by misguided energy policies and the resulting lack of investment in 80% of energy supply.

Energy starvation manifests itself with an increased risk of blackouts; this has been acknowledged across the world by reputable institutions and governments.

- In July 2022: *"The German government recommends that businesses buy emergency generators and equipment before the winter... this would be especially recommended to operators of critical infrastructure"*, FOCUS 2022.

- In June 2022, New York's grid operator NYISO 2022 wrote: *"an extreme 98-degree Fahrenheit sustained heatwave would test the system limits today and exceed grid capabilities beginning in 2023"*. In other words, extreme heat – as often happens in New York – will lead to blackouts, as reliable power capacity in the state continues to be shut down and increased intermittent capacity reaches the system.

- In April 2022, the Australian Energy Market Operator (AEMO) pointed out that: *"Australia's most populous states will face blackout risks from 2025 if new power capacity is not built in time to replace the country's biggest coal-fired plant, due to be shut that year"*, Reuters 2022b.

- In April 2022, South Africa's state power utility ESKOM warned: *"the country may have more than 100 days of electricity blackouts this year"* because of outages at its power plants (Bloomberg 2022a). According to South Africa's Integrated Resource Plan 2019, 24 GW of conventional thermal power sources (mainly coal) will be decommissioned by 2050 and replaced with "renewable" technology (SA IRP 2019).

- In August 2021, California Governor Gavin Newsom declared a state of emergency for the power grid due to concerns about supply shortages during hot summer evenings when solar production wanes and approved licenses for *"California to Build Temporary Gas Plants to Avoid Blackouts"* (Bloomberg 2021).

Oil, coal, gas, and uranium are the primary energy sources that nourish rather than starve governments, economies, and humanity. A true primary energy resource, like a true food source, does not need to be subsidized. It must, by definition, yield many times more energy (and wealth) than it consumes, or else it is a sink, not a source (see also Kiefer 2013). It is not by subsidies but rather by the merits of eROI, material efficiency, and energy density, and in spite of heavy taxation and fierce competition with other energy alternatives, that oil, coal, gas, and nuclear have grown to dominate the global energy economy with a share of over 80%.

4.3. Decarbonization and "Net-Zero"

It is evident and undisputed that (a) the world has been warming since the 1800s, the end of the Little Ice Age, (b) humans have contributed to past, present, and future global warming, and (c) airborne GHG levels contribute to global temperature change, in concert with other climate forcings. The potential effects of climatic changes may be considerable and should not be discounted, but their sources should be properly identified (see *Unsettled*, by Prof. Steven Koonin 2021).

It is also undisputed – though less known – that the global warming impact of CO_2 or any GHG declines logarithmically, thus each additional ton of any airborne GHG has less capacity to increase temperatures (Wijngaarden and Happer 2020). CO_2, after water vapor the second most important, but still a minor GHG, is active in the wavelength band from around 12-18 microns, which has been essentially saturated. This works in such a way that additional CO_2 molecules can only slightly widen the wavelength band in which the molecule can impact outgoing thermal radiation. In the absence of GHGs, the Earth's surface would be on average around −18°C; however, with GHGs, the surface reaches a temperate and livable ~15°C. The 33K difference is ascribed to the so-called "greenhouse effect" of the atmosphere (WMO 2021). We are of the opinion that our present knowledge and computational methods are far short of a predictive capability when it comes to our climatic systems.

It is not the subject of this book to quantify the causes or impacts of a warming planet – whether positive or negative. However, the discussion on "climate change" and "decarbonization" is, practically speaking, a discussion about energy and is therefore very relevant to this book. Below, we summarize a few pertinent points that are neither in dispute nor controversial, at least from an energy economic point of view:

1. ***According to the 1ˢᵗ Law of Thermodynamics, humans' heat emissions contribute to the current energy balance of the Earth.*** The vast majority of the 170.000 TWh of primary energy produced to sustain humanity ends up as heat. Global annual primary energy supply approximates the energy it would take, if theoretically channeled to do so, to melt over 1.500 km³ of ice every year (for comparison, the Arctic currently loses roughly 150 km³ of ice every year).
   The so-called "heat island effect" (see Figure 36) has been documented by Soon et al. 2015 and many other researchers. Dr. Bodo Wolf 2021 summarized this issue in detail in one of his recent German preprints, made available to Lars Schernikau. Wolf concludes (translated from German): *"Climate neutrality does not come from decarbonized economies"* since the consumed energy ends up in high-entropy, low-value heat (see Chapter 3.3) which warms our biosphere. The more inefficient our energy systems are, logically, the more we warm our planet to utilize the same amount of useful energy.
2. CO_2 and other GHG concentrations in the atmosphere are not constant and are also impacted by non-human causes, including global temperatures. For instance, the gas-holding capacity of oceans is diminished as temperatures increase, which is documented by the Vostok Ice Core data (CDIAC 2021) showing that CO_2 increases followed temperature increases prior to industrialization.
   The Quaternary is the Earth' most recent glacial period that started about 2,6 million years ago and lasts until today. It was significantly warmer prior to the Quaternary (Figure 1). **The current interglacial has lasted longer than previous interglacials** and is part of the Holocene within the Quaternary "ice age". We cannot predict how long the present interglacial will persist, but we can expect the return of glaciation to be a major shock to humanity when it arrives. Milankovitch orbital forcing has already decreased by over 40 W/m², yet the present interglacial has shown a temperature decrease of around 1°C since the broad maximum temperature of between 8.000 and 6.000 years ago (Vinther et al. 2009).
   Post-industrialization increases in atmospheric CO_2 concentrations are to a large extent due to fossil fuel combustion, and CO_2 concentrations are now over 30% higher than during the previous

four interglacials. For example, since 1984, atmospheric CO_2 concentrations increased ~20% to ~0,04% today; during the same period, atmospheric CH_4 concentrations increased ~15% to ~0,0002% today.

3. The 2018 winner of the Nobel Prize in Economic Sciences, Prof. William Nordhaus, and others calculate an economic impact of global warming of less than 5% of a multiple higher GDP in 2100, unrealistically assuming no mitigation (Nordhaus 2018, Lomborg 2020, and Kahn 2021), which should be balanced against the costs and impacts of forcing energy systems away from conventional fuels prior to having a truly sustainable energy solution, as discussed in this book. There are also many projections showing a higher economic impact of global warming.

4. The documented and undisputed *benefits of increased CO_2 concentration in the atmosphere due to its fertilization effects on plants need to be fairly evaluated* and considered (see Haverd et al. 2019 and Idso 2021 for an extensive list of peer-reviewed literature).

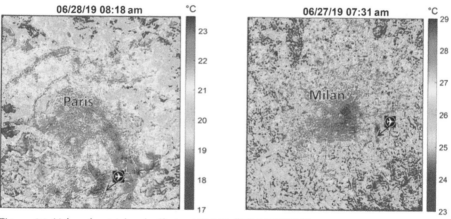

Figure 36: Urban heat island effect – NASA's "ECOSTRESS"

Note: Maps of European cities show ECOSTRESS surface temperature images from the mornings of June 27 and 28, 2019, during a heatwave. Airports and city centers are hotter than surrounding regions because they have more surfaces that retain heat (asphalt, concrete, etc.).

Source: NASA based on JPL-Caltech, (https://go.nasa.gov/3ADkPJ4)

Replacing fossil fuels with "carbon"-free energy sources by 2050 in order to reach "Net-Zero" with wind and solar PV is not realistic (Figure 37) under any scenario, as the world will be short of the energy, space, raw materials, and money required to do so. As an example, IEA Solar 2022, p62, confirmed: "*Current and planned [solar PV] manufacturing capacity is insufficient to meet the IEA Net Zero trajectory.*" The idea that such a "Net-Zero" pathway negatively impacts economies and the environment, however, currently still appears in dispute (see also Schernikau and Smith 2021 and Schernikau et al. 2022).

There are well-documented shortcomings of and disagreements among climate models on the future state of the planet's climate (Koonin 2021). Capt. K. Scott Pugh summarized correctly in 2022 that *"proven science can be modeled, but models cannot prove science"*. However, despite these large uncertainties and the inability to predict the future, to avoid even a potentially small probability of the allegedly catastrophic consequences of CO_2-induced global warming, the world has embarked on one of humanity's most drastic technological adjustments: the desired complete "decarbonization" of global energy systems within a few decades.

These "decarbonization" efforts are focused on establishing a fossil-fuel-free, and in most cases nuclear-free, energy system. The proponents of these efforts hope for completion by 2050 in order to stay on a modeled pathway known as the "1,5°C pathway". To solve the intermittency problem of wind and solar, hydrogen is the proposed solution. Having studied the energy economics behind the "energy transition", we question the economic and environmental viability and therefore desirability of such "Net-Zero" plans. Our economic arguments are independent of any concerns one may have about the impact of future global warming. The complications of moving away from conventional fuels with today's technology have been discussed in detail in this book. Only **the New Energy Revolution** detailed below may allow weaning off fossil fuels in the long run.

McKinsey 2022a confirms the **higher costs of "decarbonization" policies, especially to poorer societies**: "*Poorer countries and those reliant on fossil fuels are most exposed to the shifts in a Net-Zero transition.*" Wood Mackenzie 2022 also states that poorer nations will be hit harder than more-developed nations, having to sacrifice more relative GDP loss: *"Less developed and low-income economies will bear a disproportional burden when it comes to the cost of transition."* The study also estimates in its rather positive calculations that Africa will lose ~10% GDP per capita from the "Net-Zero" energy transition by 2050. Lomborg 2020 calculated that in the IPCC's "sustainable" SSP1 scenario, global GDP per capita in 2100 would be 30-40% less than in the IPPC's "fossil-fuel-driven" scenario SSP5. These documented additional costs of the so-called "green" energy transition, especially to poorer societies, are entirely caused by the lower eROI (summarizing net energy efficiencies) of variable "renewable" energy. Lomborg 2020 also explained that climate models and climate scenarios essentially assume no adaptation to climatic changes, which is clearly wrong and will lead to unrealistic modelled scenario outcomes.

Several studies confirm the negative economic impact of climate change mitigation on the low-income class (e.g., Fujimori et al. 2020, Lomborg 2020). Logically, the ratio of income spent on energy and food is far higher in low-income versus middle-income families. Developing nations and the poor will therefore be penalized more on a per capita basis (Figure 38). Many studies (incl. Fujimori et al. 2020), however, use climate models and often the unrealistic RCP8.5-SSP5 scenarios to show that the cost of unabated climate change would be higher to the poor. They argue that carbon taxes could be used to tackle poverty and offset price increases for energy and food products. However, the *IPCC's pathway referred to as RCP8.5 assumes per capita coal consumption in 2100 that is 6x higher than today. Thus, such scenarios may have the scientific purpose of modeling extremes but have little to do with reality*. Details are available in the highly recommended research paper *How Climate Scenarios lost Touch with Reality* (Pielke and Richie 2021a).

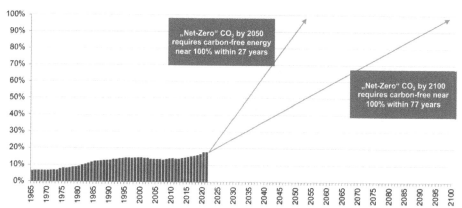

Figure 37: Global primary energy from "carbon-free" sources 1965-2100

Note: Percentage of "carbon-free" energy from wind, solar, hydro, biomass, geothermal, etc. as per BP.

Source: Pielke 2022 based on BP 2022

In order to "save the climate", investments in the energy transition away from fossil fuels have reached multiple trillion US dollars annually. BCG 2021c estimates US$ 3-5 trillion p.a., totaling US$ 100-150 trillion, will be required from now until 2050. Wood Mackenzie 2022 modeled the economic effects of a "Net-Zero" pathway and estimates a cumulative US$ 75 trillion loss by 2050. Most of the costs are borne before 2030 as "Net-Zero" actions need to be heavily front-loaded, and the benefits, as uncertain as they are, are hoped for later in the century. WoodMackenzie estimates a turning point after 2035, with GDP losses being recouped by 2100. The assumptions behind these models, however, do not fully consider the energy economics as detailed in Chapter 3 nor any cost of energy shortages and the resulting impact on human and industrial life.

McKinsey 2022a estimates a higher cost for "Net-Zero" by 2050, totaling US$ 275 trillion or US$ 9,2 trillion per year on average, an annual increase of as much as US$ 3,5 trillion from today. According to the Economist 2021, *"All efforts to 'combat' or 'act on' climate change are focused on one goal: a stable global climate, anchored by an average global temperature that no longer rises with each passing year"*, entirely dismissing well-documented natural climate variations. Since we do not truly understand all reasons why the climate has warmed and cooled over the past millennium, the assumption that the entire putative warming over the past 150 years is due to human actions does not even agree with "the climate models". The proclaimed global goal of reducing future projected temperature rises has unlocked practically unlimited financial resources – which are entirely at the expense of the taxpayer or consumer and, in our view, are being spent misguidedly.

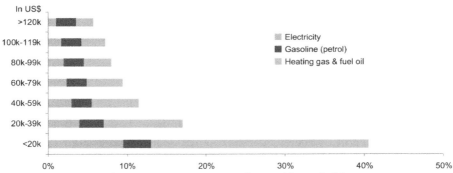

Figure 38: Household income spent on energy by total household income
Source: Eschenbach 2017

4. THE PROJECTED FUTURE OF ENERGY AND "DECARBONIZATION"

If one were to assume and accept the theory that increasing atmospheric concentrations of GHGs are the key factor or control knob to climatic changes, then it would be a fallacy to focus on human-energy CO_2 emissions – or any CO_2 tax for that matter – as discussed in Schernikau and Smith's 2022 research paper "Climate Impacts of Fossil Fuels". With IPCC and IEA data, **it can be determined that anthropogenic methane currently makes up 65% of all anthropogenic CO_2^{eq} and CO_2 accounts for the remaining ~35%**, using IPCC's 20-year global warming potential. As a result, it turns out that coal appears to be more "beneficial for the climate" than LNG because of fugitive methane emissions along the value chain. Imagine if the European leaders had known of this research result and had declared coal and nuclear to be "green" instead of gas and nuclear, as voted by the EU in July 2022 (CNBC 2022).

One conclusion from Schernikau and Smith's 2022 research is that it is evident that global GHG policies – if they cannot be avoided – should at a minimum include documented changes in *measured airborne CO_2^{eq}* (which also includes other GHGs and adjusts for over 50% natural uptake of CO_2), rather than solely CO_2 emissions measured during combustion, in order to avoid spending large amounts of public funds on ineffective or sub-optimal policies. Later, Kemfert 2022 and BP 2022 also covered the methane and gas supply chain in more detail.

Another and more pertinent conclusion is that *environmental efforts need to consider the entire value chain and life-cycle impacts to the environment from the emissions and non-emissions of our energy system* (Figure 39). Non-emission environmental impacts must consider energy input, material requirement, room costs (or space requirement), waste, water use, animal and plant life, health and safety, and more. In UMass' new study, Moran et al. 2022 detail just one of those non-emission environmental impacts in the form of water sustainability when extracting lithium in Chile. Schernikau and Smith 2022 make the point that, if CO_2 emissions need to be reduced, one of the most effective ways would be to install ultra-supercritical USC power plants, if required coupled with CCUS technology (Tramosljika et al. 2021). Before undertaking CCUS at scale, we must balance, clearly and accurately, the risks and benefits of a slightly higher atmospheric CO_2 content against the risks associated with pressurizing and burying large amounts of carbon dioxide within our Earth's crust or oceans.

It is important to remember that the reduction of human-energy CO_2 emissions – purely based on climate models – is only predicted to reduce future temperature. Through such temperature reduction, it is modeled and hoped to limit future extreme weather events and sea-level rise. In the meantime, the IPCC confirms: *"... there is low confidence in observed trends in small-scale severe weather phenomena such as hail and thunderstorms..."* and *"There is low confidence in a global-scale observed trend in drought or dryness (lack of rainfall) ..."* (IPCC 2013, pages 232 and 66). Significant scientific uncertainty remains regarding the causes and impacts of climatic changes, as well as the climate models themselves (Kooning 2021, Voosen 2021, Pielke and Richie 2021a). However, ***there is no scientific uncertainty that the financial and human costs of climatic catastrophes have been reducing on a per capita and GDP-adjusted basis***. The remaining uncertainty, however, neither precludes nor gives reason to limit serious efforts to reduce the negative environmental impacts of energy generation on our planet. Environmental degradation and climate change are not the same.

Despite the actual experienced climatic changes to date, it is evident that the modeled climate impact in 2100 is based on "average" climate models that are (a) fed scenarios that are far from reality, (b) use climate sensitivities that have proven to be too high, (c) assume the world will not adapt, (d) dismiss CO_2's undisputed fertilization effects, (e) dismiss humans' non-GHG effects, and (f) cannot explain climatic changes prior to 1850 because they largely dismiss natural variability.

We welcome and strongly support any sustainable effort that increases efficiencies and reduces the environmental strain of energy production and utilization globally. We understand that humans and GHGs contribute to slight warming, along with other climate forcings. "Renewable" energy, such as but not limited to geothermal and hydro, and storage solutions require additional research and investment to increase their energy densities and efficiencies. However, ***all energy always requires taking resources from our planet and processing them, thus negatively impacting the environment. It must be our aim to minimize these negative impacts in a meaningful way through investment, not divestment.***

5. The realistic future of energy and sustainability

Electricity is to modern civilization what blood is to the human body. Electricity is literally what makes our modern lives possible. However, do not forget that electricity makes up "only" around 40% of total energy, with heat, transport, and industry accounting for the rest (Figure 3 on page 26). Thus, governments' energy policies are of the utmost importance and have three objectives:

(1) Security of supply,
(2) Affordability of supply, and
(3) Environmental protection.

Today's energy policies, however, focus simplistically on reducing anthropogenic (human-caused-energy) CO_2 emissions with the goal of limiting or reducing future global warming (Figure 40). Therefore, today's energy policies are embarking on the almost complete "decarbonization" of energy, primarily replacing fossil fuels with net energy inefficient wind and solar.

As demonstrated by Glasgow's COP26 meeting results from November 2021, including but not limited to the *"Global Coal to Clean Power Transition Statement"* (UN-COP26 2021), many nations' energy policy decisions today pay less attention to objectives (1) and (2), and even most aspects of (3), such as plant/animal life, land/space use, material & energy input, and recycling efficiency (see Figure 8, Figure 39, and Figure 40). The 2022 Russia/Ukraine crisis has put new focus on energy security, at least in Europe, which has to a large extent relied on Russian energy raw material supply and has spent 20 years reducing its own energy independence (see Germany's political decisions to abandon coal and nuclear and the EU's extensive initiatives to divest from reliable fossil fuel energy sources). This new focus, however, seems ad hoc rather than strategic in many countries.

The objective of global investment in the "energy transition" should be to meet all three primary goals of energy policy, not just the one sub-goal of reducing human-energy CO_2 emissions.

We have explained in this book that *today's misguided energy investment focus on variable "renewable" energy increases the risk of energy starvation, with all its consequences* (see also Chapter 4.2).

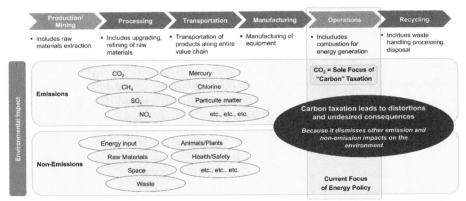

Figure 39: Environmental impact of energy systems – why carbon taxation leads to distortions and undesired consequences
Source: Schernikau Illustration

The full cost of electricity (FCOE) and eROI illustrate that wind and solar are, unfortunately, not the solution to humanity's energy problem. At grid scale, they will lead to undesired economic and environmental outcomes. The use of LCOE for the purpose of discussing the "green" energy transition must cease because it continues to mislead decision makers. Governments, industries, and educational institutions are urgently encouraged to spend additional time on learning about and discussing energy economic realities before forcing the basis of today's existence away from proven and relatively affordable energy systems. Essentially, only energy can solve the food and water crisis, only energy can enable recovery from natural disasters, and only energy can eradicate poverty. *We must do everything in our power to increase net energy efficiencies (eROI) in the production of electricity and, entirely independently and very distinctly, further increase energy efficiencies in the utilization of electricity.*

The current dramatic planned increase in installed solar and wind capacity, as detailed in Figure 34, has one advantage: It reduces the amount of fossil or nuclear fuel required, assuming no increase in power demand. However, this one advantage comes at significant cost to our environment and economies, as detailed herein. The costs to the environment originate from low energy densities, the intermittency, and inherently low eROI of variable "renewable" energy when the entire value chain and life cycle is taken into consideration (Figure 39). Energy follows different laws than computing, and technological advances will not be able to overcome the laws of physics and chemistry, for example low capacity factors or energy's tendency to increase entropy (reduce its value), in particular every time it is converted or stored. Only when we fairly compare the environmental and economic advantages and disadvantages of each energy source can we make informed energy policy decisions and prioritize future energy systems.

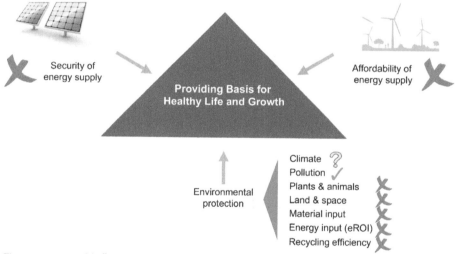

Figure 40: Variable "renewable" energy does not fulfil objectives of energy policy
Source: Schernikau Illustration

It becomes apparent that **any "carbon" taxation leads to distortions and undesired consequences because it dismisses non-"carbon" emissions and any non-emission impact** of energy systems on our environment. It also becomes apparent that investment in, not divestment from, conventional forms of energy is the only viable environmental and economic pathway until there is a so-called **New Energy Revolution** (Figure 41). The **New Energy Revolution** is a point in time when humanity may sustainably wean itself off fossil fuels. Such a new energy system may be completely new, possibly a combination of fusion or fission, solar, tidal, geothermal, or a presently unknown energy source (see also Manheimer 2022). It would likely harness the power of the nuclear force, the power of our planetary system (i.e., sun, gravity), and the energy from within our planet. It will have little to do with today's wind and photovoltaic technologies due to the physical limits of energy density, or energy available per m^2, and – most importantly – their intermittency.

As previously mentioned in Lars Schernikau's foreword, humanity has amassed more scientific knowledge since World War II than over the previous one million years of human development. Lars also pointed out that humanity's development will not be limited to scientific advances. It will also include spiritual advances that allow us to better understand the energetic connections between matter and mind, letting us access more parts of our brain and therefore find solutions we could not yet dream of. The argument is that our "energy problem" will be solved within a century or two through the **New Energy Revolution**. Unfortunately, as explained in the book, today's wind and solar technologies are not "the solution".

To achieve this **New Energy Revolution**, more must be invested in education and basic research (energy generation, material extraction & processing, storage, superconductors, etc.). Just as important is continuous investment in conventional energy to make it more efficient and environmentally friendly. It must be noted, however, that non-CO_2-emitting forms of energy generation will have no heat-offset in the form of greening and fertilizing CO_2 (see Haverd et al. 2019 and Idso 2021 for an extensive list of peer-reviewed literature). The reduced net energy efficiency of variable "renewable" energy and the increased generation of energy from non-fossil origins will logically cause an increase in low-value or high-entropy heat; this would continue to warm our surroundings even if no GHGs were emitted. In addition, byproducts of coal consumption, co-generated heat, sulfur, and fly ash would no longer be available and would need to be produced separately, requiring additional energy.

Future research and development should concentrate on understanding the true eROI of energy systems to aid prioritization (see Figure 31: Example of proposed eROI study to compare coal-fired and solar PV energy generation). Such research should also detail and quantify the full cost of electricity (FCOE) and eROI for conventional and variable "renewable" energy systems. This work requires funding, a larger team, and will be a global effort.

To further optimize conventional energy systems, future research and development should also concentrate on reducing the emission- and non-emission-related environmental impacts of existing energy systems. This should include more efficient mining and fossil fuel extraction, ultra-supercritical power plants (USC), and high-efficiency, low-emission (HELE) technologies for increasing their efficiencies, whether they are powered by fossil fuel, nuclear, or "renewable" energy. USC technology would have an immediate positive effect on nature at significantly lower costs than installing grid-scale variable "renewable" energy systems with their required backup (see also Tramosljika et al. 2021). If CO_2 emissions need to be reduced, one of the most energy- and material-efficient ways to do so would be to equip USC power plants with carbon capture utilization and storage (CCUS) technology. However, the undisputed benefits of increased CO_2 concentrations in the atmosphere, due to its photosynthetic and growth effects (fertilization) on plants, need to be considered in energy policy decisions as well.

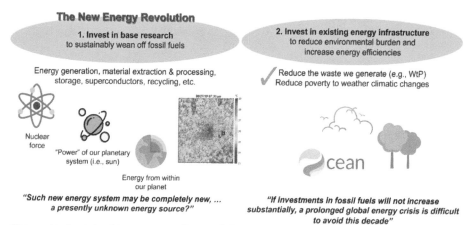

Figure 41: *Sustainable energy policy and the New Energy Revolution*
Source: Schernikau Illustration

Investment in – not divestment from – conventional energy is the logical path to not only eradicate (energy) poverty and improve the environmental and economic efficiency of fossil-fuel-installed capacity (be it for transportation, heating, or generating electricity) ***but also to avoid prolonging the energy crisis*** that started in the second half of 2021.

Executive Summary

Electricity is to modern civilization what blood is to the human body. Understanding how electricity works becomes more imperative as its importance grows significantly with the increasing electrification of transport, industry, and heating. This book is an introduction to electricity systems and electricity costs, while it also touches on primary energy and transportation. The book focuses more on the generation of electricity from a macroeconomic "energy transition" point of view and less on the details of how electricity physically works. We conclude with some thoughts on the future of energy and suggestions for energy policy, taking into account the new challenges that come with global efforts to "decarbonize".

In 2021, the debate regarding energy security (or rather electricity security) began. This debate was driven by an increase in electricity demand, a shortage of energy raw material supply, insufficient and erratic electricity generation from wind and solar, and geopolitical changes, which in turn resulted in high prices and volatility in major economies. This was witnessed around the world, including but not limited to China, Japan, India, Australia, South Africa, the US, and of course Europe. A reliable electricity supply is paramount for economic growth, which in turn leads to the eradication of poverty. What makes electricity so special and challenging to supply is that a working electricity system assumes demand equals supply at all times, that is, at every second of every day. A working electricity system requires a stable voltage and frequency so that current demand can always be met.

We examine the current drive toward a "decarbonized" world focused on wind and solar for the generation of electricity. We illustrate the gap between installed energy capacity and actual electricity generation when it comes to variable "renewable" sources of energy, such as wind and solar. Investment in oil, coal, and gas, which generate 80% of global primary energy, is less than 10% of the relative investment that wind and solar receive, which generate around 3% of global energy. What will the consequences be?

In the next few years, depending on global economic growth rates, "renewables" may only meet half the additional growth in global electricity demand or only around 20% of primary energy demand growth. Growth in electricity demand will surpass primary energy at the same time that "renewable" energy is quickly reaching physical and chemical limits. These specified limitations show that the present energy policy is largely misguided.

Realistic primary energy growth of ~50% by 2050, driven by ~25% population growth and ~20% average per capita energy demand growth, contrasts with IRENA's, McKinsey's, or the IEA's "Net-Zero" pathways, which assume much less growth and either flat or a ~10% drop in primary energy demand within eight years or by 2030.

A "hydrogen storage revolution" is hoped for in order to overcome the intermittency issues of wind and solar. However, physical and chemical principles put economic and environmental limits on grid-scale hydrogen storage adoption. Variable "renewable" energy (VRE) in the form of wind and solar has several shortcomings that even a viable storage solution would not overcome. VRE will always face a disconnect between installed capacity and generated electricity driven by:

a) low energy densities and efficiencies, resulting in large space requirements,
b) low unpredictable natural capacity factors between 5-45%, resulting in erratic and unpredictable electricity production,
c) conversion, frequency conditioning, and transmission (in)efficiencies,
d) high material input (MIPS),
e) short lifetime, and
f) recycling difficulties and economics
g) leading to low net energy efficiency (eROI).

Natural capacity factors (see preface) worldwide are a direct result of the location of the wind or solar installation; they do not in any way depend on and cannot be influenced by the technology employed. The material inefficiency of variable "renewable" energy adds to its energy inefficiency and environmental footprint. In addition, any storage – which always adds complexity and requires an energy transformation (see 2nd Law of Thermodynamics) – will always further reduce the eROI and material efficiency of an energy system because it costs or "wastes" energy. However, no grid-scale, long-term storage solution will truly solve the energy problem.

We argue that only primary energy can determine reasonably well the total energy input for producing, using, and recycling our (energy) infrastructure.

It must be humanity's aim to increase energy generation efficiencies and material efficiencies and therefore increase eROI and reduce or at least optimize material input. Wind, solar, and biofuels have an average eROI below the required threshold to sustain modern civilization and are therefore strongly advised against at grid scale.

Additional concerns about raw material supply and processing are entering the international political arena. China dominates almost all "green" raw material processing supply chains, including critical solar PV, wind, and EV production. Energy security is about the supply of energy raw materials – namely oil, coal, gas, and uranium – and now also about who controls the supply of short-lifetime "renewable" energy production capacity and consuming equipment.

It is prudent to assume that wind and solar alone will not be able to generate enough electricity to match the expected total energy demand, and it would be inadvisable to force its grid-scale adoption. It goes without saying that any loss of "renewable" energy due to conversion for storage or transportation is inefficient, will contribute to warming our biosphere, and has to be avoided at all costs.

In summary:
1. Future energy requirements outstrip "Net-Zero" pathways and possible "renewable" generation.
2. "Renewable" energy remains a scarce resource for the foreseeable future, and its widespread adoption would reduce humanity's net energy efficiency below the level required to sustain civilization at its present living standard and population.
3. No viable, long-term, grid-scale storage solution has yet been found or proposed, and any storage solution would not truly solve the energy problem.

Levelized cost of electricity (LCOE) is a marginal cost measure and is inadequate to compare intermittent forms of energy generation with dispatchable ones when making decisions at a national or societal level. Using full cost of electricity (FCOE), which defines the full cost to society, wind and solar are not cheaper than conventional power generation and in fact, become more expensive as their penetration of the energy system increases. This is illustrated by the extraordinarily high cost of the "green" energy transition worldwide, exemplified by Germany.

Energy policy and investors should not favor wind, solar, biomass, geothermal, hydro, nuclear, gas, or coal but should support all energy systems in a manner which avoids energy shortages and energy poverty, including variable "renewable" energy systems where they make sense. If investments in fossil and nuclear sources of energy do not increase substantially and soon, it will be difficult to avoid a prolonged global energy crisis. If CO_2 emissions need to be reduced, one of the most effective ways would be to couple ultra-supercritical, high-efficiency, low-emission (USC HELE) power plants with CCUS technology.

CCUS also costs energy and therefore reduces eROI, but it would be environmentally and economically more stable and efficient than installing variable "renewable" energy systems at grid scale. Further, the benefits of CO_2 through fertilization need to be fairly evaluated and considered together with an unemotional comparison of the costs of future climatic changes versus the costs of transitioning to lower-carbon-emitting energy systems with today's available technologies. These costs are always higher for poorer societies because of the resulting rising costs of electricity.

All energy consumption requires taking resources from our planet and processing them, thus negatively impacting the environment. It must be humanity's goal to minimize negative impacts in a meaningful way through investment rather than divestment, and increasing rather than decreasing energy and material efficiencies.

The modeled climate impact in 2100 is based on "average" climate models that are (a) fed scenarios that are far from reality, (b) use climate sensitivities which have proven to be too high, (c) assume the world will not adapt, (d) dismiss CO_2's undisputed fertilization effects, (e) dismiss humans' non-GHG effects, and (f) cannot explain climatic changes prior to 1850 because they largely dismiss natural variability. The potential effects of climatic changes may be considerable and should not be discounted, but their sources should be properly identified (see *Unsettled*, by Prof. Steven Koonin 2021).

Therefore, we urge energy policymakers to refocus on energy policy's three objectives: energy security, energy affordability, and environmental protection. This translates into two paths for the future of energy:

(1) Invest in education and basic research to pave the way toward a **New Energy Revolution** in which energy systems can sustainably be weaned off fossil fuels, but at an increase in energy availability per capita and increase of net energy efficiency (eROI).

(2) In parallel, energy policy must **support investment in conventional energy** systems to improve their efficiencies and reduce the

environmental burden of energy generation for our lives, at least until the New Energy Revolution has been realized.

We must do everything in our power to increase the net energy efficiencies (eROI) of our energy generation systems, and – independently and distinctly – further increase energy efficiencies in the consumption of energy. The production and consumption of energy are often considered together, but they are in fact entirely different. This book focuses on the production or generation of electricity, not its consumption.

Additional research is required to better understand eROI, the total cost of energy, material input, and the effects of current energy transition pathways on global energy security.

Keywords

Energy policy, electricity, fossil fuels, natural gas, coal, nuclear, wind, solar, renewable, energy, capacity factor, cost of electricity, eROI, energy returns, energy efficiency, clean coal technology, USC, carbon capture, CCUS.

References

All references where last accessed 25 June 2022 and are mostly buffered in case website content changes.

AGE 2021, AG Energiebilanzen e.V. 2021, Zusatzinformationen, (https://bit.ly/3RAdaSJ).

Air Products 2021, ACWA Power and NEOM Sign Agreement for $5 Billion Production Facility, July 2020, (https://bit.ly/3Rql9QS).

Allwood, Julian and Jonathan Cullen 2015, Sustainable Materials without the hot air, UIT Cambridge, Sep 2015, Book ISBN 978-1-906 860-30-1, (https://bit.ly/3zU11BE).

Agora 2022, Agora Energiewende, Energiewende Deutschland Stand 2021, (https://bit.ly/3cEBZhg), for German electricity data during April and May 2022 (https://bit.ly/3ehuQUP).

Atasu et al. 2021, INSEAD, The Hidden Cost of Solar Energy, Dec 2021, (https://bit.ly/3q42Giq); also published by Harvard Business Review, HBR, Jun 2021, (https://bit.ly/3wPfkWq).

Barron-Gafford et al. 2016, The Photovoltaic Heat Island Effect: Larger Solar Power Plants Increase Local Temperatures, Scientific Reports 6, no. 1, (https://bit.ly/3clYvWq).

Baruya 2019, former IEA-Clean Coal Center: The Economic and Strategic Value of Coal, CCC/296, (https://bit.ly/3B7SHPH).

Bauer et al. 2021, authored by Bauer, Thomas, Christian Odenthal, and Alexander Bonk, Molten Salt Storage for Power Generation, Chemie Ingenieur Technik 93, no. 4 (January 2021): 534-46, (https://bit.ly/3ALvt07).

BCG and IEF 2020, Oil and Gas Investment in the New Risk Environment (https://on.bcg.com/3ekZnAM); International Energy Forum (IEF' is world's largest international organization of energy ministers from 70 countries.

BCG 2021a, BCG CEO Guide to Cyber Security, Sep 2021, (https://on.bcg.com/3RxkZZ6).

BCG 2021b, Mastering Scale in Renewables, Jun 2021, (https://on.bcg.com/3QaCYU8).

BCG 2021c: Executive Perspectives: The time for climate action is now, (https://on.bcg.com/3TAi1oG).

Bleich, Daniel 2022, Eine Einführung in das Energienetz – Wieso uns Wind und Sonne aktuell nicht retten und es auch (noch) keine Speicher gibt, Ruhrbarone, Aug 2022, (https://bit.ly/3CnEEER).

BfWE 2020, Bundesministerium für Wirtschaft und Energie, EEG in Zahlen: Vergütung, Differenzkosten und EEG-Umlage 2000-2021, (https://bit.ly/3wPGBbq).

Bloomberg 2021, By Mark Chediak and Naureen S Malik, California to Build Temporary Gas Plants to Avoid Blackouts, Aug 2021, (https://bloom.bg/3B8i5Vr).

Bloomberg 2022a, South Africa Is Poised for 101 Days of Power Outages This Year, Apr 2022, (https://bit.ly/3ekcAdq).

Bloomberg 2022, by Shankleman, Alberto Nardelli, Chiara Albanese and Jessica, Big Blow to Climate Change as Germany Pushes for G7 Reversal of Fossil Fuel Commitments, 25 June 2022, without paywall (https://bit.ly/3RdI6YZ).

BNEF 2021, Bloomberg New Energy Finance: Energy Transition Investment Trends, (https://bit.ly/3CS9HdT).

Bossel, Ulf and B. Eliasson 2006, Energy and the Hydrogen Economy, Proceedings of the IEEE, Vol. 94, No. 10, Oct 2006, 36, (https://bit.ly/3Rc7LBo).

Bossel et al. 2009, Ulf Bossel, Baldur Eliasson, and Gordon Taylor, The Future of the Hydrogen Economy: Bright or Bleak?, Cogeneration and Distributed Generation Journal, Volume 18, Issue 3, Dec 2009, (https://bit.ly/3Rc7LBo).

BP 2019, BP Statistical Review of World Energy, Primary Energy, (https://on.bp.com/3Rxjzy0); BP donates TPES/capita at 75,7GJ/capita or 21.023 kWh/capita in 2019.

BP 2020, BP Statistical Review of World Energy 2020 (https://on.bp.com/3KHA6NI).

BP 2021, BP Statistical Review of World Energy 2021, (https://on.bp.com/3D8J2dd).

BP 2022, BP Statistical Review of World Energy 2022, (https://on.bp.com/3Rabcs9).

Caraballo et al. 2021, authored by Caraballo, Adrián, Santos Galán-Casado, Ángel Caballero, and Sara Serena, Molten Salts for Sensible Thermal Energy Storage: A Review and an Energy Performance Analysis, Energies 14, no. 4 (January 2021): 1197, (https://bit.ly/3ebABmJ).

Cairney-RenewEconomy 2021, authored by Professors Cairney, Hutchinson, Preuss, and Dr. Chen, Embrittlement: Hydrogen Could Be the Future of Energy – but There's One Big Road Block, Mar 2021, (https://bit.ly/3wU5SkD).

Carbajales-Dale et al. 2014, Can We Afford Storage? A Dynamic Net Energy Analysis of Renewable Electricity Generation Supported by Energy Storage, Energy & Environmental Science 7, no. 5, (https://bit.ly/3KGnJ4c).

CDIAC 2021, Historical Carbon Dioxide Record from the Vostok Ice Core, Dec 2021, (https://bit.ly/3eevAtG).

CEA 2020, Central Electricity Authority, Indian coal space requirement from CEA, April 2020, (https://bit.ly/3RvI5PX).

Chen et al. 2019, The Potential of Photovoltaics to Power the Belt and Road Initiative, Joule 3, (https://bit.ly/3cHakfK).

CleanTechnica 2022, by Michael Barnard, Energy Vault Loses $1.2 Billion/40% Market Cap, CO_2e/KWh Worse Than Natural Gas, CleanTechnica, Apr 2022. (https://bit.ly/3cKjLef).

CNBC 2022, authored by Clifford, Catherine, Europe Will Count Natural Gas and Nuclear as Green Energy in Some Circumstances, CNBC, Jul 2022, (https://cnb.cx/3CSbVd9).

Collins, Leigh 2022a, World's Largest Green Hydrogen Project Unveiled in Texas, with Plan to Produce Clean Rocket Fuel for Elon Musk | Recharge, RechargeNews, Mar 2022, (https://bit.ly/3TCDqgN).

Collins, Leigh 2022b, 'Europe Is Never Going to Be Capable of Producing Its Own Hydrogen in Sufficient Quantities': EU Climate Chief, RechargeNews, Latest renewable energy news, May 2022, (https://bit.ly/3B4I6F8).

Columbia 2022, SIPA Center on Global Energy Policy, Hydrogen: A Hot Commodity Lacking Sufficient Statistics, by Anne-Sophie Corbeau, Apr 2022, (https://bit.ly/3ecksgV).

Davie, Michael 2022, The Rush for Cobalt in the Congo Reveals the Human Cost of the World's Green Energy Future – ABC News, Feb 2022, (https://bit.ly/3RiCw7W).

Delannoy et al. 2021, EROI – peak oil and the low-carbon energy transition: A net-energy perspective, Applied Energy 304, Aug 2021, 117 843, (https://bit.ly/3cKmCUv).

DeSantis et al. 2022, authored by DeSantis, James, Houchins, Saur, and Lyubowsky, Cost of long-distance energy transmission by different carriers, iScience, Dec 2021, (https://bit.ly/3DymVey).

Dierig, Carsten 2022, die Welt, Energiekosten: 200.000 Jobs in Gefahr – Stahlindustrie im Klima-Dilemma, 2022, (https://bit.ly/3wMrr6E).

DOE 2015, Department of Energy, Quadrennial Technology Review 2015, p390 Table 10.4, Sep 2015, (https://bit.ly/3ejBDgD).

Derwent et al. 2006, authored by Derwent, R., P. Simmonds, S. O'Doherty, A. Manning, W. Collins, C. Johnson, M. I. Sanderson, D. Stevenson, Global Environmental Impacts of the Hydrogen Economy, 2006, (https://bit.ly/3AI35fA).

Deyfuss et al. 2022, authored by Dreyfus, Xu, Shindell, Zaelke, and Ramanathan, Mitigating Climate Disruption in Time: A Self-Consistent Approach for Avoiding Both Near-Term and Long-Term Global Warming, Proceedings of the National Academy of Sciences 119, no. 22, May 2022, (https://bit.ly/3AI35fA).

DW 2022, Deutsche Welle (www.dw.com), 'Grüner Stahl' – welche Erfolgschancen hat er wirklich?, 10.06.2022, DW.COM, Jun 2022, (https://bit.ly/3wQeIQq).

EC 2022, EC Subsidies – Energy Taxation, Carbon Pricing and Energy Subsidies, Jan 2022, (https://bit.ly/3cEIPTY).

Economist 2021, The Economist, The Climate Issue: Hitting "Net-Zero" by 2050, Newsletter 22 Mar 2021, (https://bit.ly/3ALmaNO).

EIA 2021, US Energy Information Administration, Annual Energy Outlook 2021, Feb 2021, (https://bit.ly/3wSVNV7).

EIA AEO Assumptions 2022, Assumptions on the Annual Energy Outlook 2022: Electricity Market Module, Mar 2022, (https://bit.ly/3QpZeK5).

EPA 2022, Energy Protection Agency of the USA: Economics of Biofuels, Overviews and Factsheets, Apr 2022, (https://bit.ly/3q3dxJA).

ERCOT 2021, Statesman, Historic winter storm freezes Texas wind turbines hampering electric generation, (https://bit.ly/3CRpPwh).

EuroAsia Interconnector 2017, The EuroAsia Interconnector document, Oct 2017, (https://bit.ly/3Qgh69P).

Eschenbach, Willie 2017, The Cruelest Tax Of All | Skating Under The Ice [wordpress.com] (https://bit.ly/3epSzCk), Detail: the Energy Information Agency (EIA) collects relevant data (https://bit.ly/3RAoyh8), with exception of gasoline usage. Eschenbach collected most recent data (https://bit.ly/3wSYcyW) for 2009 (Excel workbook). Gasoline usage figures are here (https://bit.ly/3B66Xaz) from the US Bureau of Labor Statistics. Income averages by tiers are available here (https://bit.ly/3em7fSE) from the US Census Bureau.

Everett, Bruce 2021, The Social Cost of Carbon and Carbon Taxes – Pick a Number, Any Number, CO_2 Coalition, (https://bit.ly/3ALOm38).

FOCUS 2022, Bundesregierung rät Firmen zum Kauf von Notstromaggregaten, FOCUS Online, Jul 2022, (https://bit.ly/3B7kiQU).

Fraunhofer 2022, Fraunhofer Institute, Energy Charts, (https://bit.ly/3eieUkS).

Fujimuri et al. 2020, An assessment of the potential of using carbon tax revenue to tackle poverty, Environ. Res. Lett. 15 114 063, (https://bit.ly/3eiCyxS).

Gaertner, Edgar, 2022, „Erneuerbare" versauen die Netzfrequenz und verursachen Kurzschlüsse, Sep 2022, (https://bit.ly/3CjRLao).

Gilder, George 2018, Life After Google: The Fall of Big Data and the Rise of the Blockchain Economy, Book ISBN 9 781 621 575 764, Chapter 8, (https://amzn.to/3APwGE2).

Global Wind Atlas 2022, Wind map, (https://bit.ly/3B94ltw).

Global Solar Atlas 2022, Solar map, (https://bit.ly/3APwWmu).

Greenberg, Andy 2019, Sandworm – a new era of cyberwar and the hunt for the Kremlin's most dangerous hackers, Nov 2019, Book ISBN 978-0-385-54 440-5, (https://bit.ly/3CSAXcu).

GSU 2021, Georgia State University, on heat pumps, Dec 2021, (https://bit.ly/3Qq1YXT).

H_2-View 2021, $9.4bn H_2 Megaproject Set for Namibia, Nov 2021, (https://bit.ly/3wQjUUq).

Harvey, Fiona 2021, The Guardian, May 2021, (https://bit.ly/3B7KieS).

Haverd et al. 2019, Higher than Expected CO_2 Fertilization Inferred from Leaf to Global Observations, Global Change Biology 26, no. 4 (Nov. 2019): 2390-2402, (https://bit.ly/3efd9VJ).

HC Group 2021, Hydrogen: Hope, Hype and Thermodynamics with Paul Martin, Oct 2021, (https://bit.ly/3Ryt7se).

Hoskins 2022, Ed Hoskins Blog and calculations, edmhdotme, The Excess Costs of European Weather-Dependent Power Generation 6/2022, Jun 2022, (https://bit.ly/3eihveC).

IAEA PRIS 2022, The International Atomic Energy Agency – Power Reactor Information System, detailed Statistics, Jul 2022, (https://bit.ly/3RrTDUV).

Idso, Craig 2021, extensive list of peer-reviewed academic papers confirming CO_2 fertilization, CO_2 Science, (https://bit.ly/3wOWwqc & https://bit.ly/3cExgfx).

IEA 2019, Is Exponential Growth of Solar PV the Obvious Conclusion? by Brent Wanner, Feb 2019, (https://bit.ly/3q1CDst).

IEA Power 2019, Installed power generation capacity in the Stated Policies Scenario, Nov 2019, (https://bit.ly/3KFBZKu).

IEA Statistics 2019, Statistics Questionnaire FAQ, Nov 2019, (https://bit.ly/3Rcu5uJ).

IEA Investments 2020, World Energy Investments 2020, (https://bit.ly/3ALGyOL).

IEA WEO 2020, World Energy Outlook 2020, (https://bit.ly/3B791zX).

IEA Electricity 2021, Projected Costs of Generating Electricity 2020, Dec 2020, (https://bit.ly/3ADceWG).

IEA Statistics 2021, Key World Energy Statistics, Sep 2021, (https://bit.ly/3CSj3pV).

IEA Investments 2021, World Energy Investments 2021, (https://bit.ly/3RtNeYB).

IEA Electricity 2021, Electricity Market Report, Jul 2021, (https://bit.ly/3qb4J4g).

IEA WEO 2021, World Energy Outlook 2021, (https://bit.ly/3B8AcKU).

IEA Net-Zero 2021, Net Zero by 2050 – Analysis, May 202, (https://bit.ly/3B7LH58).

IEA Minerals 2021, The Role of Critical Minerals in Clean Energy Transitions, May 2021, (https://bit.ly/3RdW3WP).

IEA 2022, Energy transitions require innovation in power system planning, Jan 2022, (https://bit.ly/3KGeYXP).

IEA Electricity 2022, Electricity market report, Jan 2022, (https://bit.ly/3wNtd7K).

IEA WEO 2022, World Energy Investments 2022, (https://bit.ly/3CSpGbO).

IEA Solar 2022, Solar Report – Supply Chains – The World Needs More Diverse Solar Panel Supply Chains to Ensure a Secure Transition to Net Zero Emissions, (https://bit.ly/3RtzuNF).

IEEJ 2020, Institute of Energy Economics Japan, IEEJ Energy Outlook 2020, Jan 2020, (https://bit.ly/3KHK6Gn).

IEEJ 2021, Institute of Energy Economics Japan, IEEJ Energy Outlook 2021, Oct 2020, (https://bit.ly/3cKsyNh).

IMF 2021, A Global and Country Update of Fossil Fuel Subsidies, Sep 2021, (https://bit.ly/3Ba0b4O).

Investopeida 2020, on learning curve, (https://bit.ly/3cEPb5M).

IPCC 2013, IPCC Assessment Report 5, Working Group 1, Climate Change 2013: The Physical Science Basis, (https://bit.ly/3RcvjWR).

IRENA 2020, Energy Subsidies – Evolution in the Global Energy Transformation to 2050, (https://bit.ly/3QbPD9l).

IRENA 2021, World Energy Transitions Outlook: 1.5°C Pathway, Jun 2021, (https://bit.ly/3Rsp2WL).

IRENA 2022, World Energy Transitions Outlook: 1.5°C Pathway, Mar 2022, (https://bit.ly/3wQ2iZ0).

IRENA LCOE 2022, LCOE – Renewable Power Generation Costs in 2021, July 2022, (https://bit.ly/3RqJ43G).

JP Morgan 2022, Annual Energy Paper – The Elephants in the Room, May 2022, (https://bit.ly/3TB3h8Y).

Kahn et al. 2021, Long-Term Macroeconomic Effects of Climate Change" Energy Economics 104, (https://bit.ly/3D3ZZoK).

Kalt et al. 2022, authored by Kalt, Thunshirn, Krausmann, and Haberl, Material Requirements of Global Electricity Sector Pathways to 2050 and Associated Greenhouse Gas Emissions, Journal of Cleaner Production 358, Apr 2022): 132014, (https://bit.ly/3AGphqw).

Kearney 2021, Refining 2021: Who Will Be in the Game?, (https://bit.ly/3cKrCsw).

Kemfert, Claudia 2021, Translated from Prof. Kemfert in RND, Wasserstoff ist nicht das neue Öl, Sep 2021, (https://bit.ly/3cEiQMu).

Kemfert et al. 2022, authored by Claudia, Fabian Präger, Isabell Braunger, Franziska M. Hoffart, and Hanna Brauers, The Expansion of Natural Gas Infrastructure Puts Energy Transitions at Risk, Nature Energy, Jul 2022, 1–6, (https://bit.ly/3cJOM27).

Kiefer, Capt. Ike 2013, Twenty-First Century Snake Oil, Feb 2013, (https://bit.ly/3Q7oEM9), see page 6 in Chapter 3.3 on hydrogen and page 15 in Chapter 5 on eROI.

Kis et al. 2018, Electricity Generation Technologies: Comparison of Materials Use, Energy Return on Investment, Jobs Creation and CO_2 Emissions Reduction, Energy Policy 120, (https://bit.ly/3QggYr8).

Kleijn et al. 2011, authored by Kleijn, René, Ester van der Voet, Gert Jan Kramer, Lauran van Oers, and Coen van der Giesen. "Kleijn et al 2011: Metal Requirements of Low-Carbon Power Generation", Energy 36, no. 9, Sep 2011, (https://bit.ly/3RNF1yX).

Kober et al. 2020, Global energy perspectives to 2060 – WEC's World Energy Scenarios 2019, Energy Strategy Reviews 31: 100 523, (https://bit.ly/3qf3tgN).

Koonin, Steven 2021, Book: – nsettled – What Climate Science Tells Us, What It Doesn't, and Why It Matters. BenBella Books, Book ISBN 978-1-950 665-79-2, (https://bit.ly/3wXAfH0).

KU Leuven 2022, Europe's Metal Association, Metals for Clean Energy: Pathways to Solving Europe's Raw Materials Challenge, April 2022. (https://bit.ly/3Qb4HEk).

Lane, Charles 2022, Washington Post Article, Review of Vaclav Smil's 2022 book, A scientist's inconvenient truths about decarbonizing the economy, May 2022, (https://wapo.st/3RDRCV2).

Lomborg, Bjorn 2020, Welfare in the 21st century, Technological Forecasting & Social Change 156 (2020) 119 981, (https://bit.ly/3RdMWFE).

Lu et al. 2020, Impacts of Large-Scale Sahara Solar Farms on Global Climate and Vegetation Cover, Geophysical Research Letters 48, e2020GL090 789, (https://bit.ly/3RG0xpg).

Manheimer, Wallace 2022, Civilization Needs Sustainable Energy – Fusion Breeding May Be Best, Journal of Sustainable Development 15, no. 2: p98, (https://bit.ly/3ASaWXU).

MAN Energy Solutions 2021, MAN Molten Salt Energy Storage Report (MOSAS), Nov 2021, (https://bit.ly/3RTXDxh).

Mar et al. 2022, authored by Mar, Kathleen A., Charlotte Unger, Ludmila Walderdorff, and Tim Butler, Beyond CO_2 Equivalence: The Impacts of Methane on Climate, Ecosystems, and Health, Environmental Science & Policy 134, Apr 2022: 127-36, (https://bit.ly/3BfaeFK).

McFadden, Christopher 2019, The Kardashev Scale: How Far Can Humanity Advance as a Civilization?, Feb 2019, (https://bit.ly/3Rm2aZs).

McKinsey 2018, Automation and the Workforce of the Future, May 2018, (https://mck.co/3TIgJbf).

McKinsey 2021a, Global Energy Perspectives 2021, (https://mck.co/3ASCs7w), and 2019, (https://mck.co/3wTO9tO).

McKinsey 2021b, Net-Zero Power: Long-Duration Energy Storage for a Renewable Grid, Nov 2021, (https://mck.co/3esSg9S).

McKinsey 2022a, The Net-Zero Transition: Its Cost and Benefits, Jan 2022, (https://mck.co/3cNCd5S).

McKinsey 2022b, Christer Tryggestad et al., Global Energy Perspective 2022 Executive Summary, Apr 2022, (https://mck.co/3qhsqlc).

Mearns, E. 2016, ERoEI for Beginners, Energy Matters (blog), (https://bit.ly/3CY6vNP).

Michaux, Simon 2021, Finland, Assessment of Size and Scope of Non Fossil Fuel System to Phase out Oil Gas and Coal, Sep 2021, (https://bit.ly/3BfFrbV).

Miller, L. and Keith, D. 2018, Climatic Impacts of Wind Power, Joule 2, (https://bit.ly/3wYmqIa).

MISO 2021, US Midcontinent Independent System Operator, MISO Report Concludes Upwards of 50% Renewables Is Achievable with Careful Planning, Renewable Energy World (blog), Feb 2021, (https://bit.ly/3CYNJG4).

Montel 2022, Norway to Curb Power Exports on Hydro Dearth, August 2022, (https://bit.ly/3Rkuk6S)

Moore, Michael 2020, Movie: Planet of the Humans | Full Documentary | Directed by Jeff Gibbs, (https://bit.ly/3CYIlmD).

Moore, Patrick 2017, The Positive Impact of Human CO_2 Emissions on the Survival of Life on Earth, Electronic Book Publisher, March 2017, (https://bit.ly/3BeeyVs).

Moran et al. 2022, authored by Moran, Boutt, McKnight, Jenckes, Munk, Corkran, and Kirshen, Relic Groundwater and Prolonged Drought Confound Interpretations of Water Sustainability and Lithium Extraction in Arid Lands, Earth's Future 10, no. 7, Jul 2022, (https://bit.ly/3RDtq58).

Morris et al. 2019, A manganese hydride molecular sieve for practical hydrogen storage under ambient conditions, Energy & Environmental Science, Issue 5, 2019, (https://rsc.li/3wX6VQU).

Murphy, L. and Elimä, N. 2021, In Broad Daylight – Uyghur Forced Labour in the Solar Supply Chain, Sheffield Hallam University, May 2021, (https://bit.ly/3cNDPMY).

Nahle, Nasif 2009, Cycles of Global Climate Change, Biology Cabinet Journal Online, Jul 2009, (https://bit.ly/3cJPyMH); referencing C.R. Scotese, Analysis of the Temperature Oscillations in Geological Eras, 2002; W.F. Ruddiman, Earth's Climate: Past and Future, New York, NY: W.H. Freeman and Co., 2001; Mark Pagani et al., "Marked Decline in Atmospheric Carbon Dioxide Concentrations during the Paleocene", Science 309, no. 5734 (2005): 600-603.

NASA 2019, NASA Earth Observatory 2019, Online 2021, (https://go.nasa.gov/3QfB5Wb).

NDR3 2020, Nord Deutscher Rundfunk – Extra 3, Ein Gaskraftwerk in Frührente, Mar 2019, (https://bit.ly/3qbvRQR and https://bit.ly/3D6DUWC).

Netzfrequenz.info, Aktuelle Anzeige der Netzfrequenz in Deutschland mit Skala über den maximal möglichen Frequenzbereich und Erklärungen, Sep 2022, (https://bit.ly/3fSMCyj).

Nguyen et al. 2021a, Long-Term Quantification and Characterisation of Wind Farm Noise Amplitude Modulation, Measurement 182 (Sep 2021): 109 678, (https://bit.ly/3erurzf).

Nguyen et al. 2021b, Benchmark Characterisation and Automated Detection of Wind Farm Noise Amplitude Modulation, Applied Acoustics 183 (Dec 2021): 108 286, (https://bit.ly/3wTQyok).

Nordhaus, Ted 2018, Projections and Uncertainties about Climate Change, American Economic Journal, Vol. 10, No. 3, (https://bit.ly/3FRLhTn).

NU 2020, Northwestern University, Gas storage method could help next-generation clean energy vehicles, (https://bit.ly/3KKDOWI).

NYISO 2022, Empire Center for Public Policy. NYISO: New York Electric Grid Remains at Risk, Jun 2022, (https://bit.ly/3cHKwQQ).

OECD NEA (2018), The Full Costs of Electricity Provision, OECD Publishing, Paris, (https://bit.ly/3AP6qJV).

OurWorldInData 2020, Energy Mix, Nov 2020, (https://bit.ly/3QcPAKp).

OurWorldInData 2021, World Population Growth, Our World in Data based on HYDE, UN, and UN Population Division (2019 Revision), (https://bit.ly/3ARVpY4).

Paul, Sonali 2022, World's First Hydrogen Tanker to Ship Test Cargo to Japan from Australia, Reuters, Jan 2022, sec. Environment, (https://reut.rs/3Rxad52).

Pielke, R. and Ritchie, J. 2021a, How Climate Scenarios Lost Touch With Reality, Jul 2021, (https://bit.ly/3BfdG3a).

Pielke, R. and Richie, J. 2021b, Distorting the View of Our Climate Future, Energy Res. & Social Sc. 72: 101 890, (https://bit.ly/3wW3oSG).

Pielke, Roger 2022, Global Decarbonization: How Are We Doing, Substack newsletter, The Honest Broker by Roger Pielke Jr. (blog), July 2022. (https://bit.ly/3AOUbNI).

Pitt, Charles H. and Wadsworth, Milton, E. 1981, Current Energy Requirements in the Copper Producing Industries, JOM 33, no. 6 (1981): 25-34, (https://bit.ly/3q9UvBl).

Polimeni et al. 2015, The Myth of Resource Efficiency: The Jevons Paradox, Book ISBN 978-1-138-42297-1, (https://amzn.to/3RISfmw).

Prieto, P. and Hall, C. 2013, Introduction: Solar Energy and Human Civilization, New York, NY, Springer, Book ISBN 978-1-4419-9437-0, (https://bit.ly/3ARWjnq).

pv-education 2020, refining silicon, accessed 4 Oct 2020, (https://bit.ly/3QfRs5m), and Bine Information Services, accessed 4 Sep 2020, (https://bit.ly/3Rx4qwk).

pv-Magazine 2022, pv-magazine International, World Has Installed 1TW of Solar Capacity, Mar 2022, (https://bit.ly/3cQLOcg).

Rankin, John 2012, Energy Use in Metal Production, High Temperature Processing Symposium, Swinburne University of Technology, 2012, (https://bit.ly/3KO91In).

REN21 2021, Renewables Global Status Report, (https://bit.ly/3x0Hn5I).

Recharge News 2022a, Liquid Hydrogen as Shipping Fuel, (https://bit.ly/3q6LskI).

Recharge News 2022b, Hydrogen twice as powerful a GHG a previously thought, Hydrogen GWP_{20}, (https://bit.ly/3AMD8vv).

Reuters 2022a, authored by Chestney, Nina, and Vera Eckert, Analysis: Running Short of Gas: Russia's Pipeline Repair Has Europe Worried, Reuters, 24 Jun 2022, sec. Energy, (https://reut.rs/3KK76EV).

Reuters 2022b, by Paul Sonali, Australia's Eastern States Face Blackout Risk from 2025, Reuters, April 2022, sec. Asia Pacific, (https://reut.rs/3qf8jut).

Roos, W. and Vahl, C. 2021, Infraschall aus technischen Anlagen – ASU, Jul 2021, (https://bit.ly/3CXSShL).

Ruhnau, O. and Qvist, S. 2022, Storage Requirements in 100% Ren. Elec. System: Extreme Events and Inter-Annual Variability, Environmental Research Letters 17, no. 4, Mar 2022, (https://bit.ly/3RB7987).

Smith, W. and Schernikau, L. 2022, An introduction to wind and wind turbines, unpublished draft paper, Apr 2022, (https://bit.ly/3RzfsBo).

SA IRP 2019, South Africa Integrated Resource Plan 2019, Oct 2019, (https://bit.ly/3wVQAfc).

Sanicola, Laura 2022a, reported on Reuters, Global Refiners Falter in Efforts to Keep up with Demand, May 2022, sec. Commodities, (https://reut.rs/3qejd3q).

Sanicola, Laura 2022b, reported on Reuters, Reuters: Explainer: Why is there a worldwide oil-refining crunch?, June 2022, sec. Commodities, (https://reut.rs/3RxmQNA).

Schernikau, L. and Smith, W. 2021, Solar in Spain to Power Germany, SSRN Electronic Journal, Apr 2021, (https://bit.ly/3RARKVw).

Schernikau, L. and Smith, W. 2022, Climate impacts of fossil fuels in today's electricity systems, SAIMM Vol. 122, No. 3, Mar 2022, (SSRN https://bit.ly/3RdTc06 and https://bit.ly/3CZyltd).

Schernikau et al. 2022, Full cost of electricity 'FCOE' and energy returns 'eROI', Journal of Management and Sustainability, Canadian Center of Science and Education, Apr 2022, (https://bit.ly/3RyCp7O).

ScienceDirect 2011, ScienceDirect, Turbidite, Deep-Sea Sediments, (https://bit.ly/3RU4Xcf).

Smil, Vaclav 2016, Energy Transitions: Global and National Perspectives, (https://bit.ly/3RjOfmn).

Smil, Vaclav 2017, Energy and Civilization: A History, Book ISBN 0 262 035 774, (https://bit.ly/3q9A73t).

Smil, Vaclav 2022, How the World Really Works: The Science Behind How We Got Here and Where We're Going, Viking, Book ISBN 978-0-593-29 706-3, (https://amzn.to/3TJcFHz).

S&P 2021, S&P Platts, Coal Trader International, 29 Dec 2021.

Soon et al. 2015, Re-evaluating the role of solar variability on Northern Hemisphere temperature trends, Earth-Science Reviews 150, 409-52, (https://bit.ly/3KMFMpo).

Stein, Ronald 2022, Cfact, Energy Shortages and Inflation the New Norm as Refinery Closures Outpace Construction, June 2022, (https://bit.ly/3KOcNBx).

Supekar, S. and Skerlos, S. 2015, Reassessing the Efficiency Penalty from Carbon Capture in Coal-Fired Power Plants, Environmental Science & Technology 49, (https://bit.ly/3cQPf2E).

UN 2019, population data, (https://bit.ly/3eoOYEd).

Toke 2021, Options for energy storage, 100% Renewable UK, Oct 2021, (https://bit.ly/3RDB3Zy).

Tramosljika et al. 2021, Advanced Ultra-Supercritical Coal-Fired Power Plant with Post-Combustion Carbon Capture, Sustainability 12(2), 801, (https://bit.ly/3RynoU5).

UN-COP26 2021, UN Climate Change Conference (COP26), Glasgow 2021, Global Coal to Clean Power Transition Statement, Nov 2021, (https://bit.ly/3TL5RJy).

University of Texas 2018, Full Cost of Electricity, Energy Institute, The University of Texas at Austin, April 2018, (https://bit.ly/3wYuUz0).

Vinther at al. 2009, Holocene thinning of the Greenland ice sheet, Nature 461, Sep 2009, (https://go.nature.com/3RyHnRW).

Voosen 2021, IPCC Confronts Implausibly Hot Forecasts of Future Warming, Science, AAAS, Jul 2021, (https://bit.ly/3BggH2U).

Watter, Holger 2021, Wasserstoff aus Wind, (https://bit.ly/3REZ9CW).

WEF 2019, World Economic Forum 2019, video on vegetation cover (https://bit.ly/3RgTbZu).

WEF 2020, World Economic Forum, Charts on global energy supply, Sep 2020, (https://bit.ly/3RyHyN6).

Weissbach et al. 2013, Energy intensities, EROIs, and energy payback, Energy 52, p210-221, (https://bit.ly/3QfWw9S).

Wijngaarden, W. A. and Happer, W. 2020, Dependence of Earth's Thermal Radiation on Five Most Abundant GHGs, arXiv:2006.03098 [physics], (https://bit.ly/3REZYf0).

Willuhn, Marian 2022, pv-magazine, Satellite Cyber Attack Paralyzes 11 GW of German Wind Turbines, (https://bit.ly/3qcAbPK).

WMO 2021, World Meteorological Organization, Greenhouse gases, (https://bit.ly/3x0LK0f).

Wolf, Bodo 2021, Der Beitrag der Menschen zur Erwärmung der Biosphäre sind nicht Kohlendioxid, sondern Wärmeemissionen, Dec 2021.

WoodMackenzie 2022, No Pain, No Gain – the Economic Consequences of Accelerating the Energy Transition, Jan 2022, (https://bit.ly/3TL6DWY).

WU Vienna 2020, Material flows by material group, 1970-2017. Visualization based upon the UN IRP Global Material Flows Database, Vienna University of Economics and Business, (https://bit.ly/3QhISTq).

Zalk, J. and Behrens, P. 2018, The Spatial Extent of Renewable and Non-Renewable Power Generation, Energy Policy 123, (https://bit.ly/3BfxV0u).

Zhu et al. 2016, Greening of the Earth and Its Drivers, Nature Climate Change 6, no. 8 (August 2016): 791-95, (https://bit.ly/3RlZjQ4).

Appendix 1

Thermodynamic systems and energy units

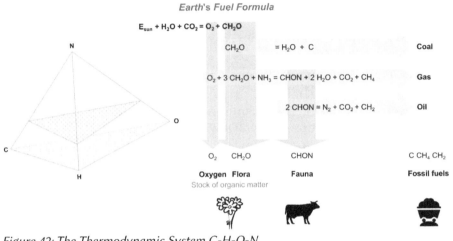

Figure 42: The Thermodynamic System C-H-O-N
Source: Wolf 2021

Input \ Output	kJ	kcal	kWh
1 kJ	1	0,2388	0,000278
1 kcal	4,1868	1	0,001163
1 kWh	3.600	860	1
1 kg ce	29.308	7.000	8,14
1 kg oe	41.868	10.000	11,63
1 m³ natural gas	31.736	7.580	8,816
1 BTU	1,05506	0,252165392	0,000293072

Figure 44: Conversion of Energy Units

APPENDIX 1

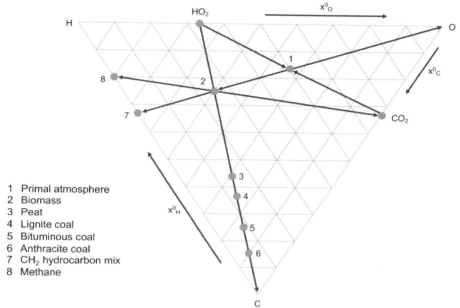

1 Primal atmosphere
2 Biomass
3 Peat
4 Lignite coal
5 Bituminous coal
6 Anthracite coal
7 CH_2 hydrocarbon mix
8 Methane

Figure 43: Fuels in a Thermodynamic System $C-H_2-O_2$; Conversion of Carbohydrate to Coal, Methane and Liquid Hydrocarbons

Source: Wolf 2021

Kg ce	kg oe	m³ natural gas	BTU
0,000034	0,000024	0,000032	0,947817
0,000143	0,0001	0,00013	3,96566633
0,123	0,086	0,113	3.412,14
1	0,70	0,924	27.778,13
1,428	1	1,319	39.683,05
1,082	0,758	1	-
3,599954E-5	2,519968E-5	-	1

Appendix 2

Schernikau et al. 2022, Full cost of electricity 'FCOE' and energy returns 'eROI'

The peer-reviewed scientific paper Schernikau et al. 2022, authored by Dr. Lars Schernikau, Prof. William Smith, and Prof. Rosemary Falcon "Full Cost of Electricity 'FCOE' and Energy Returns 'EROI'" discusses the full cost of electricity and energy returns. The paper was published in May 2022 in the Canadian Journal of Management and Sustainability 12, no. 1, p96.

https://doi.org/10.2139/ssrn.4000800

APPENDIX 2

By Dr. Lars Schernikau, Prof. William Hayden Smith, Prof. Rosemary Falcon (Vers. 08/2022)

Full cost of electricity 'FCOE' and energy returns 'eROI'

Abstract

Understanding electricity generation's true cost is paramount to choosing and prioritizing our future energy systems. This paper introduces the full cost of electricity (FCOE) and discusses energy returns (eROI). The authors conclude with suggestions for energy policy considering the new challenges that come with global efforts to "decarbonize".

In 2021, debate started to occur regarding energy security (or rather electricity security) which was driven by an increase in electricity demand, shortage of energy raw material supply, insufficient electricity generation from wind and solar, and geopolitical challenges, which in turn resulted in high prices and volatility in major economies. This was witnessed around the world, for instance in China, India, the US, and of course Europe. Reliable electricity supply is crucial for social and economic stability and growth which in turn leads to eradication of poverty.

We explain and quantify the gap between installed energy capacity and actual electricity generation when it comes to variable renewable energy. The main challenge for wind and solar are its intermittency and low energy density, and as a result practically every wind mill or solar panel requires either a backup or storage which adds to system costs.

LCOE is inadequate to compare intermittent forms of energy generation with dispatchable ones and when making decisions at a country or society level. We introduce and describe the methodology for determining the full cost of electricity (FCOE) or the full cost to society. FCOE explains why wind and solar are not cheaper than conventional fuels and in fact become more expensive the higher their penetration in the energy system. The IEA confirms "...*the system value of variable renewables such as wind and solar decreases as their share in the power supply increases*". This is illustrated by the high cost of the "green" energy transition.

We conclude with suggestions for a revised energy policy. Energy policy and investors should not favor wind, solar, biomass, geothermal, hydro, nuclear, gas, or coal but should support all energy systems in a manner which avoids energy shortage and energy poverty. All energy always requires taking resources from our planet and processing them, thus negatively impacting the environment. It must be humanity's goal to minimize these negative impacts in a meaningful way through investments – not divestments – by increasing, not decreasing, energy and material efficiencies.

Therefore, the authors suggest energy policy makers to refocus on the three objectives, energy security, energy affordability, and environmental protection. This translates into two pathways for the future of energy:

(1) invest in education and base research to pave the path towards a **New Energy Revolution** where energy systems can sustainably wean off fossil fuels.

(2) In parallel, energy policy must support **investment in conventional energy** systems to improve their efficiencies and reduce the environmental burden of generating the energy required for our lives.

Additional research is required to better understand eROI, true cost of energy, material input, and effects of current energy transition pathways on global energy security.

About the authors:

Dr. Lars Schernikau is an energy economist, entrepreneur, and commodity trader, Zurich, Switzerland

Prof. William Hayden Smith is Professor of Earth and Planetary Sciences at McDonnell Center for Space Sciences at Washington University, St. Louis, MO, USA.

Prof. Emeritus Rosemary Falcon is recently retired DSI-NRF SARChl Professor from the Engineering Faculty at the University of the Witwatersrand, Johannesburg, South Africa.

Available online at *SSRN*

Reprinted and adapted from Schernikau et al. 2022 Full cost of electricity 'FCOE' and energy returns 'eROI', Journal of Management and Sustainability Vol. 12, No. 1, June 2022 issue at *Canadian Center of Science and Education*.

Acknowledgement

The authors declare that they did not receive any financial or other support from any company or organization for this paper.

Selected Abbreviations

CCUS – Carbon capture utilization and storage
eROI – Energy return in energy invested
VRE – Variable renewable energy, such as wind and solar
HELE – High efficiency, low emission
IEA – International Energy Agency in Paris
FCOE – Full cost of electricity
LCOE – Levelized cost of electricity
PES – Primary energy supply or PE for primary energy
PV – Photovoltaic
USC – Ultra-super-critical
VRE – Variable renewable energy
~ – Approximately

Download this document

YouTube Video "The Future of Energy"

147

Appendix 3

Schernikau and Smith 2022, Climate Impacts of Fossil Fuels in Modern Energy Systems

The peer-reviewed scientific paper Schernikau and Smith 2022 "Climate impacts of fossil fuels in modern energy systems" compares coal and gas climate impacts drawing solely on IPCC and IEA data. The paper was published in March 2022 at SAIMM, Journal of the Southern African Institute of Mining and Metallurgy, vol. 122, no. 3, pp. 133-146, 122, no. 3.

https://papers.ssrn.com/abstract=3968359

APPENDIX 3

„Climate Impacts" of Fossil Fuels in Today's Electricity Systems

Abstract

Oil, coal, and gas account for ~80% of global PE, but only a portion of total airborne CO_2^{eq} (~40% at GWP_{20} to ~60% at GWP_{100}), even though they account for 95% of total measured CO_2 emissions. Benefits of these energy sources as well as their related costs are not all incorporated in current energy policy discussions. Global GHG policies must include documented changes in measured airborne CO_2^{eq} to avoid spending large amounts of public funds on ineffective or sub-optimal policies.

The authors examined airborne CO_2, which is less than half of emitted CO_2, as well as reported methane emissions and the global warming potential (GWP) of methane as published by the IPCC for coal and natural gas. The authors point out their reservations about IPCC's GWP*. The surprising conclusion is that surface-mined coal appears "better for the climate" than the average natural gas and all coal appears beneficial over LNG. Therefore, current CO_2-only reduction policies and CO_2 taxes are leading to misguided consequences and the switch from coal to gas, especially LNG, will not have the desired impact of reducing predicted future global warming, quite the contrary. Approximately 30% of global warming since pre-industrial times is attributed to methane by the IPCC and IEA, while it must be noted that methane emissions from natural sources account for ~40% and agriculture for ~25% of annual global methane emissions. Energy accounts for ~20% of documented methane emissions.

This study finds that CO_2 contributes only ~35% of annual airborne anthropogenic GHG emissions after accounting for methane, over a 20-year horizon. At a 100-year horizon, CO_2's contribution increases to ~60%. Energy policy that does not consider all GHG emissions along the entire value chain will lead to undesired economic and environmental distortions. All carbon taxation and CO_2 pricing schemes are incorrect and need to be revised.

At IPCC's 20-year GWP, a ~2% higher loss of methane[1] across the value chain prior to combustion of natural gas versus coal will lead to "climate parity" of coal with natural gas. Analyzing public data, natural gas value chains have high methane and undocumented CO_2 losses. On global average, using only IEA documented methane data, natural gas emits ~15% more CO_2^{eq} than surface-mined coal over a 20-year horizon. This difference increases as the use of shale gas and LNG expands.

Investors should support all reliable energy systems in a manner which avoids an energy crisis and allows human development and eradication of poverty. If CO_2 emissions need to be reduced, one of the most effective ways would be to install ultra-super critical power plants with CCUS technology. However, the undisputed benefits of increased CO_2 concentrations in the atmosphere because of its photosynthetic and growth effects (fertilization) on plants need to be considered in energy policy decisions as well. The authors suggest that future research and development should concentrate on reducing net emissions of fossil fuel power plants and providing cost effective and reliable conventional new power generation, utilizing clean coal and clean natural gas technology.

About the authors:
Dr. Lars Schernikau is an energy economist, investor and commodity trader (Orcid 6489-2117).
Prof. William Hayden Smith is Professor of Earth and Planetary Sciences at McDonnell Center for Space Sciences at Washington University, St. Louis, MO USA (Orcid 9819-7923)

Available online at SSRN and ResearchGate. Short YouTube video summary of this paper is available at https://youtu.be/9Cu4NiMu0jw0

Reprint from Schernikau, L. and Smith, W.H., 2022. Climate impacts of fossil fuels in today's electricity systems. Journal of the Southern African Institute of Mining and Metallurgy (SAIMM), vol. 122, no. 3, pp. 133-146. http://dx.doi.org/10.17159/2411-9717/1874/2022

Acknowledgement
The authors declare that they did not receive any financial or other support from any company or organization for this paper.

Key Abbreviations
CH_4	– Methane
CCUS	– Carbon Capture Utilization and Storage
CO_2	– Carbon Dioxide
CO_2^{eq}	– CO_2 equivalents incorporating CO_2 and CH_4, using IPCC's Global Warming Potential
GHG	– Greenhouse Gas
Gt	– Giga tons, or billion metric tons
GWP	– Global Warming Potential as defined by IPCC and specified over 20 or 100 years
IEA	– International Energy Agency in Paris
IPCC	– UN's Intergovernmental Panel on Climate Change
LNG	– Liquified Natural Gas
Mt	– Mega tons, or million metric tons
NG	– Natural Gas
PE	– Primary Energy
PNG	– Pipeline Natural Gas
USC	– Ultra-Super Critical
~	– approximately

Appendix 4

Smith and Schernikau 2022, An Introduction to Wind Energy

The scientific paper Smith and Schernikau 2022 "An Introduction to Wind Energy – Can "Renewables" Replace Fossil Fuel and Nuclear Energy in Germany?" discusses what it would take to power Germany with 100% wind. This is a theoretical calculation and illustrates the effort required. The paper also gives a detailed economic summary of the resource wind and the physical and natural boundaries that modern wind turbines cannot overcome independent of the technology employed. The paper was first made public in Elsevier's SSRN electronic journal in July 2022.

https://ssrn.com/abstract=4096843

APPENDIX 4

By Dr. Lars Schernikau and Prof. William Hayden Smith. (Vers. xx/2022)

An Introduction to Wind Energy
Can „Renewables" Replace Fossil Fuel and Nuclear Energy in Germany?

By Prof. William Hayden Smith and Dr. Lars Schernikau

Abstract

Electricity can, in principle, power all the energy demands of civilization, provided sufficient electricity can be produced reliably and at an affordable cost.

Here, we detail and summarize the physical mechanisms of wind energy, and pose the question whether the wind resource or the wind plus the solar resource can meet the entire energy demand of Germany as planned under NetZero 2050.

Our back-of-envelope, but physical analysis using only published data, shows that the attainable renewable energy supply is unlikely to match the German future energy demand. The reason is that the total land/sea area required for wind farms and PV parks to power Germany reliably approximates the area of Germany. Grid scale wind-solar energy system creates climatological effects comparable to CO_2, and threatens severe ecological consequences, some of which could be irreversible.

Mining, processing, and fabrication during the energy transition plus the rapid and continual "repowering" of the massive wind farms and PV parks results in materials and commodities demand far exceeding those anticipated. Combined with a decade of decreasing growth of wind power, "Net-Zero" 2050 goals will not be achieved without an enhanced, exponential growth.

The magnitude of the backup for German variable renewable energy is computed and compared with Norway's total hydroelectric energy storage. The Norwegian hydroelectric storage can meet German electrical demand for less than 6 days, not the required minimum one month. The growth of future electrical energy demand, under "Net-Zero" 2050, necessary to cover all energy demand, eliminates the possibility of a sufficient hydroelectric backup. Other backup options fail, leaving a fundamental unsolved issue.

The conclusion is that the 'Energiewende' leaves Germany in thrall to growing energy deficits which can be avoided only by conventional energy resources until a replacement energy supply that 'adds up' is created.

Disclaimer: This is a back-of-the-envelope calculation based on published, referenced data. The calculations depend on the stated assumptions. The interested reader may revise these BOEs with their own assumptions.

Keywords: energy policy, electricity, wind energy, wind turbines, wind, solar, renewable energy, capacity factor, cost of electricity, energy restoration

About the authors:

Prof. William Hayden Smith is Professor of Earth and Planetary Sciences at McDonnell Center for Space Sciences at Washington University, St. Louis, MO, USA. https://orcid.org/0000-0001-9839-7023

Dr. Lars Schernikau is an energy economist, entrepreneur, and commodity trader in the energy raw materials industry; Zurich, Switzerland and Singapore. https://orcid.org/0000-0001-6469-0117

Published xxx

Publicly available at

Selected Abbreviations
BOE   – Back of the envelope
CCUS – Carbon capture utilization and storage
CF   – Capacity factor
CSP   – Concentrated solar power
DR   – Dynamic reserve
ERR   – Energy extraction and restoration rate (also called energy rate of restoration)
eROI – Energy return on energy invested (also often referred to as eROEI)
EV   – Electric vehicle
GH   – 'Green' hydrogen
GHG – Greenhouse gas
GW   – Gigawatt, 1 000 MW or 10⁹ watts (power unit). 1 TW, terawatt = 1 000 GW
HELE – High efficiency, low emission
HFNT – hydrocarbon fueled and nuclear thermal power plants
IEA   – International Energy Agency in Paris
MTOe – Million tons of oil, equivalent
PA   – Peak-to-average power ratio
PE   – Primary energy (PES = primary energy supply)
PV   – Photovoltaic
T-C   – Tri-cellular circulation of the atmosphere
TWh – Terawatt-hour, 1 000 GWh or 10¹² Watt-hours, PWh, petawatt-hour = 1 000 TWh
VRE   – Variable renewable energy, such as wind and solar
WT   – Wind turbine
~   – Approximately

Appendix 5

Schernikau and Smith 2021, Solar in Spain to Power Germany

The scientific paper Schernikau and Smith 2021 "How Many km² of Solar Panels in Spain and how much battery backup would it take to power Germany" discusses what it would take to power Germany with 100% solar produced in Spain backed up by batteries. This is a theoretical calculation and illustrates the effort required. The paper was published in Elsevier's SSRN electronic journal in April 2021.

https://ssrn.com/abstract=3730155

APPENDIX 5

How many km² of solar panels in Spain and how much battery backup would it take to power Germany

1. Abstract

Written by

Dr. Lars Schernikau and
Prof. William H. Smith

Published November 2020,
last updated March 2021
(the authors appreciated all
received feedback leading
to this substantially revised
version)
Publicly available at
Researchgate & SSRN

About the authors:
Dr. Lars Schernikau is an
energy economist and
entrepreneur in the energy
raw material industry
Prof. William Hayden Smith
is Professor of Earth and
Planetary Sciences at
McDonnell Center for Space
Sciences at Washington
University, St. Louis, MO,
USA.

DOI: 10.2139/ssrn.3730155

Germany is responsible for about 2% of global annual CO_2 emissions from energy. To match Germany's electricity demand (or over 15% of EU's electricity demand) solely from solar photovoltaic panels located in Spain, about 7% of Spain would have to be covered with solar panels (~35.000 km²). Spain is the best-situated country in Europe for solar power, better in fact than India or (South) East Asia. The required Spanish solar park (PV-Spain) will have a total installed capacity of 2.000 GWp or almost 3x the 2020 installed solar capacity worldwide of 715 GW. In addition, backup storage capacity totaling about 45 TWh would be required. To produce sufficient storage capacity from batteries using today's leading technology would require the full output of 900 Tesla Gigafactories working at full capacity for one year, not counting the replacement of batteries every 20 years. For the entire European Union's electricity demand, 6 times as much – about 40% of Spain (~200.000 km²) – would be required, coupled with a battery capacity 6x higher.

To keep the Solar Park functioning just for Germany, PV panels would need to be replaced every 15 years, translating to an annual silicon requirement for the panels reaching close to 10% of current global production capacity (~135% for one-time setup). The silver requirement for modern PV panels powering Germany would translate to 30% of the annual global silver production (~450% for one-time setup). For the EU, essentially the entire annual global silicon production and 3x the annual global silver production would be required for replacement only.

There are currently not enough raw materials available for a battery backup. A 14-day battery storage solution for Germany would exceed the 2020 global battery production by a factor of 4 to 5x. To produce the required batteries for Germany alone (or over 15% of EU's electricity demand) would require mining, transportation and processing of 0,4-0,8 billion tons of raw materials every year (7 to 13 billion tons for one-time setup), and 6x more for Europe. The raw materials required include lithium, copper, cobalt, nickel, graphite, rare earths & bauxite, coal, and iron ore for aluminum and steel. The 2020 global production of lithium, graphite anodes, cobalt or nickel would not nearly suffice by a multiple factor to produce the batteries for Germany alone.

It appears that solar's low energy density, high raw material input and low energy-Return-On-energy-Invested (eROeI) as well as large storage requirements make today's solar technology an environmentally and economically unviable choice to replace conventional power at large scale.

Disclaimer: This is a back-of-the-envelope calculation based on experience from existing PV plants in California and publicly available battery data and insolation data for Spain. The calculations can be adjusted using different assumptions.

Preface:

• **Power** (in Watts, in German "Leistung") is the horsepower of a car's engine. Energy to drive a Tesla, for example, is derived from a battery. A Tesla S half-ton kWh battery powers a 192 kW electric motor to accelerate the 2,2-ton Tesla S.

• **Energy** (in Watt-hour or Wh, in German "Arbeit oder Energie") is how much work it takes to move the 2,2-ton car, for instance, for 1h at 100 km/h over a specified terrain. Energy is equivalent to "work". In this case, energy varies with travel time, velocity, mass, aerodynamics, friction, and the power applied to overcome those "obstacles". The half-ton Tesla S battery stores energy of 85 to 100 kWh.

• **Capacity Factor "CF"** (in German "Nutzungsgrad") is the percentage of power output achieved from the installed capacity for a given site, usually stated on an annual basis.

  - Capacity factor is not the efficiency factor. Efficiency measures the percentage of input energy transformed to usable energy.

  - Capacity factor assumes a stable photovoltaic response and measures the site quality, which varies with latitude, air mass, season, diurnal (24h sun cycle at that location), and weather.

  - In Germany, photovoltaics ("PV") achieve an average annual capacity factor of ~10%, while California reaches an annual average CF of 25%³. Thus, California yields 2,5x the output of an identical PV plant in Germany.

  - It is important to distinguish between the average annual capacity factor and the monthly or better weekly and daily capacity factor, which is very relevant when we try to use solar for our daily power needs (See Figure 5).

153